AutoUni – Schriftenreihe

Band 64

AutoUni – Schriftenreihe

Band 64

Herausgegeben von
Volkswagen Aktiengesellschaft
AutoUni

Emine Bostanci

Performance Analysis
of Brushless DC Machines
with Axially Displaceable Rotor

Logos Verlag Berlin

AutoUni – Schriftenreihe

Herausgegeben von
Volkswagen Aktiengesellschaft
AutoUni
Brieffach 1231
38436 Wolfsburg
Tel.: +49 (0) 5361 - 896-2104
Fax: +49 (0) 5361 - 896-2009

http://www.autouni.de

Bibliografische Information der Deutschen Nationalbibliothek

Die Deutsche Nationalbibliothek verzeichnet diese Publikation in der
Deutschen Nationalbibliografie; detaillierte bibliografische Daten sind
im Internet über http://dnb.d-nb.de abrufbar.

Zugleich: Dissertation, Leibniz Universität Hannover, 2014

ISBN 978-3-8325-3743-2
ISSN 1867-3635

Logos Verlag Berlin GmbH
Comeniushof, Gubener Str. 47,
10243 Berlin

Tel.: +49 (0) 30 / 42 85 10 90
Fax: +49 (0) 30 / 42 85 10 92
http://www.logos-verlag.de

Performance Analysis
of Brushless DC Machines
with Axially Displaceable Rotor

Von der Fakultät für Elektrotechnik und Informatik

der Gottfried Wilhelm Leibniz Universität Hannover

zur Erlangung des akademischen Grades

Doktor-Ingenieurin (Dr.-Ing.)

genehmigte Dissertation

von

M.Sc. Emine Bostanci

geboren am 03. Februar 1984 in Ankara, Türkei

2014

1. Referent: Prof. Dr.-Ing. Bernd Ponick
2. Referent: Prof. Dr.-Ing. Markus Henke

Tag der Promotion: 28. Mai 2014

Acknowledgments

This dissertation has been conducted during my employment at the department of Powertrain Systems in Group Research of Volkswagen AG as a PhD candidate.

My special thanks go to my advisor Professor Dr.-Ing. Bernd Ponick for his valuable technical supervision, interest on my topic, understanding and patience. I appreciate that he accepted to supervise this thesis despite his heavy schedule. I would like to thank Professor Dr.-Ing. Markus Henke not only for accepting to examine this study but also for his trust on me, technical support, encouragement and valuable advices through out my master's and Ph.D. studies. I am very greatfull that he gave me the chance to be a part of his team where I learned a lot and enjoyed my time. I would also like to thank Professor Dr.-Ing. Axel Mertens for accepting to be the chairmann of the examination community.

I can not thank enough to Dr.-Ing. Zdeno Neuschl for his support, encouragement, willingness to help, patience and gentle manners. It was a pleasure to work with him and to know him. Many thanks are also given to my colleagues for their support, especially Dr.-Ing. Robert Plikat for his assistance and Dr.-Ing. Christian Mertens for his advices. I must mention that I could not conduct the experiments without Oliver Rauch and Henning Strauß, thank you very much for your valuable help. I would also like to thank David Straube, Kurt-Klaus Bernhardt, Karsten Muschall and Daniel Jahns for their dedicated help in building experimental setups and in conducting experiments. Additionally, my special thanks go to my office mates, Mario Richter and Marcus Kruse, for the pleasant time together, to Dr.-Ing. Elisabeth Schulze for her friendship and to Dr.-Ing. Levent Sarioglu for his dedicated support. I would also like to thank Felix Madauß for his contribution to this study with his excellent student project and master's thesis as well as Liu Yang for his valuable work as an intern.

I would like to express my very great appreciation to Dr.-Ing. Leon Voss from ANSYS Germany GmbH for his excellent software support and valuable advices. I would also like to extend my thanks to the past and present members of IAL. I really enjoyed the time together.

My sincerely thanks go to my friends for their understanding, encouragement and the pleasant time together.

Finally, I wish to thank my family for their support, encouragement and love throughout my study. Despite the distance between, my sister was always there for me, my brother made my life enjoyable, my mother always believed in me and my father was my well beloved supervisor.

Abstract

In this study, a mechanical field weakening method that is applied to enhance the high-speed performance of a PM brushless machine by reducing the active axial machine length is analyzed in terms of its operational characteristics and efficiency. For this purpose, an existing multi-phase brushless DC (BLDC) machine with an axially displaceable rotor is used. First, the parameters of this machine, such as back EMF, torque constant and phase inductances are determined for various axial rotor positions by using 2-D and 3-D FEM analyses. Then, the determined machine parameters are used to build a dynamic simulation model of the drive system. By using this developed system model, the operational characteristics including the limits of the mechanical field weakening method are analyzed. In the second place, the efficiency of the BLDC machine is investigated, laying the main focus of attention on the additional losses in the field weakening range. In order to determine these losses, new numerical and measurement methods are developed. Furthermore, the measurement results are used to validate the numerical results. Finally, recommendations for PM brushless machines with an axially displaceable stator/rotor including possible design measures to reduce the additional losses and the application of mechanical field weakening are presented.

Key words: mechanical field weakening, axially displaceable rotor, multi-phase brushless DC machine

Kurzfassung

In dieser Studie wird ein mechanisches Feldschwächverfahren zur Verbesserung des Hochgeschwindigkeitsverhaltens einer permanentmagneterregten elektrischen Maschine durch Reduzierung der aktiven axialen Maschinenlänge im Hinblick auf Betriebseigenschaften und Effizienz analysiert. Hierzu wird eine mehrphasige bürstenlose Gleichstrommaschine mit einem axial verschiebbaren Rotor verwendet. Zunächst werden Maschinengrößen wie Gegen-EMK, Drehmomentkonstante und Phaseninduktivitäten für verschiedene axiale Rotorpositionen mittels 2-D- und 3-D-FEM-Analysen bestimmt. Mithilfe dieser Maschinenparameter wird ein dynamisches Simulationsmodell des Antriebssystems erstellt. Anhand dieses Modells werden die Betriebseigenschaften einschließlich der Drehzahlgrenzen des mechanischen Feldschwächverfahrens analysiert. Anschließend wird die Effizienz der bürstenlosen Gleichstrommaschine unter besonderer Berücksichtigung der Zusatzverluste im mechanischen Feldschwächbereich untersucht. Zur Bestimmung dieser Verluste werden neue numerische und messtechnische Methoden entwickelt und die numerischen Ergebnisse anhand der Messergebnisse verifiziert. Zum Schluss werden Empfehlungen für permanentmagneterregte elektrische Maschinen mit axial verschiebbarem Stator/Rotor einschließlich Maßnahmen zur Reduzierung der Zusatzverluste und den Anwendungsbereich der untersuchten mechanischen Feldschwächung vorgestellt.

Schlagworte: mechanische Feldschwächung, axial verschiebbarer Rotor, mehrphasige bürstenlose Gleichstrommaschine

Contents

List of Symbols

Definitions

a	Time-dependent variable
$\vec{a}$	Vector
$\hat{a}$	Peak value of a

Symbols

A	Area
$\vec{A}$	Magnetic vector potential
b	Width
$\vec{B}, B$	Magnetic flux density
B_{m}	Variable that depends on previous material magnetization
B_{mag}	Magnitude of magnetic flux density
B_{n}	Normal component of magnetic flux density
B_{r}	Remanent magnetization
B_{rad}	Radial component of magnetic flux density
C_{dc}	DC-link capacitor
d	Diameter
e, e_{p}	Induced voltage, back EMF
e_{norm}	Normalized back EMF
f	Frequency
h	Height
$\vec{H}, H$	Magnetic field intensity
H_{e}	External magnetic field
I, i	Current
i	Variable
I_{dc}	Supply DC current
i_{d}	Phase direct-axis current
i_{p}	Phase current
i_{q}	Phase quadrature-axis current
J	Moment of inertia
k	Magnetic coupling coefficient
k_{c}	Coefficient of classical eddy current losses
k_{e}	Coefficient of excess eddy current losses
k_{h}	Coefficient of hysteresis losses
k_{lam}	Stacking factor
$k\Phi$	Torque constant
L	Inductance
l	Length
l_0	Axial length of lamination

L_{app}	Apparent inductance
l_{ag}	Axial gap between rotor parts
l_{fe}	Axial length of stator stack
L_{diff}	Differential inductance
L_e	End winding leakage inductance
L_g	Air-gap leakage inductance
l_i	Ideal axial length
L_p	Phase inductance
L_s	Slot leakage inductance
L_d	Phase direct-axis inductance
L_q	Phase quadrature-axis inductance
l_{s-r}	Axial stator/rotor overlap length
M	Mutual inductance
n	Rotational speed
N	Number of phases
n_c	Number of coils in a slot
n_e	Number of evaluation points
n_p	Number of pole pairs
P	Power
P_c, p_c	Classical eddy current losses
P_{cu}	Copper losses
P_{fe}	Core losses
P_h, p_h	Hysteresis losses
P_e, p_e	Excess eddy current losses
r	Radius
R	Resistance
R_{ac}	AC resistance
R_{dc}	DC resistance
R_m	Reluctance
$R_{m,0}$	Reluctance of vacuum
$R_{m,fe}$	Reluctance of iron
R_p	Phase resistance
S	Number of slots
T	Torque
T	Electrical period
T_0	Reference temperature
T_{av}	Time average of torque
T_{em}	Electromagnetic torque
T_L	Load torque
U, u	Voltage
U_{DC}	DC-link voltage
u_d	Phase direct-axis voltage
u_p	Phase voltage
u_q	Phase quadrature-axis voltage

V	Volume
W_L	Magnetic energy stored in an inductance
W_Φ	Magnetic energy
x, y, z	Coordinates of Cartesian coordinate system
α	Angle
α_0	Linear temperature coefficient of copper
β	Exponent of hysteresis losses
δ_{skin}	Skin depth
θ	Angle
θ_{fw}	Freewheeling angle
θ_m	Rotor angular displacement
θ_{on}	Turn-on angle
θ_{off}	Turn-off angle
θ_r	Pole pair related rotor angular displacement
λ	Thermal conductivity
μ	Magnetic permeability
μ_0	Magnetic permeability of free space
μ_r	Relative magnetic permeability
μ_r^*	Equivalent relative magnetic permeability
ρ	Electrical resistivity, volumetric mass density
σ	Electrical conductivity
τ_{cp}	Diameter of an end turn
Φ	Magnetic flux
Ψ	Flux linkage
Ψ_{PM}	Flux linkage due to permanent magnet flux
Ψ_s	Flux linkage due to stator phase currents
ω	Angular frequency
ω_m	Rotor angular speed
ω_r	Pole pair related rotor angular speed

Abbreviations

ADR	Axially displaceable rotor
BLAC	Brushless alternating current
BLDC	Brushless direct current
DSPM	Doubly salient permanent magnet
EV	Electric vehicle
FCV	Fuel cell vehicle
FEM	Finite element method
FW	Field weakening
HEV	Hybrid electric vehicle
IM	Induction machine
IPM	Interior permanent magnet
PHEV	Plug-in hybrid electric vehicle

PM	Permanent magnet
PMSM	Permanent magnet synchronous machine
PAM	Pulse amplitude modulation
PWM	Pulse width modulation
SM	Synchronous machine
SMPM	Surface mounted permanent magnet

Indices

0	Free space
1, 2,...10	Phases of a 10-phase electric machine
1-2	Between 1 and 2
a, b, c	Phases of a 3-phase electric machine
add	Additional
av	Average
com	Combined
cu	Copper
elec	Electrical
ew	End winding
fe	Iron
fw	Field weakening
g	Air-gap
in	Inner
k, l	Variable
lim	Limit
max	Maximum value
mech	Mechanical
min	Minimum value
out	Outer
p	Phase
peak	Peak value
prox	Proximity effect
r	Rotor
reg	Regular
ra	Rated value
rad	Radial
rms	Root mean square
skin	Skin effect
s	Stator core stack
tan	Tangential
x	Variable
xy	In x-y plane
yz	In y-z plane

1 Introduction

The development of powertrains for passenger cars is challenging due to the demand for high vehicle performance with low exhaust emissions and low energy consumption. Electric vehicles (EVs), hybrid electric vehicles (HEVs), plug-in hybrid electric vehicles (PHEVs) and fuel cell vehicles (FCVs) are promising technologies in order to reach these design goals. Therefore, the electrification of the vehicle powertrain is an important field of research and development for car manufacturers. In EVs, HEVs, PHEVs and FCVs, electrical drives are used as generators and traction machines either alone or in combination with internal combustion engines. In other words, electrical drives are one of the key technologies for vehicles with electrified powertrains. Therefore, their development according to the requirements of automotive applications is of great importance.

Three electric machine topologies, i.e. induction machines, synchronous machines and permanent magnet (PM) brushless machines, are implemented in conventional vehicles. Among these, PM brushless machines are the choice of many auto manufacturers as reported in [1] and [2]. These machines have a higher power/torque density and a better efficiency at full load compared to other topologies due to the constant permanent magnet excitation. On the other hand, the constant permanent magnet excitation may result in a limited speed range and a worse fault tolerance capability. Therefore, a proper machine design is required to overcome these drawbacks of PM electric machines without affecting their power/torque density and efficiency.

The development of PM brushless machines with new features in order to enhance their performance is one of the current research topics. One of the novel machine designs is the PM brushless machine with a variable active machine length. The active machine length is reduced at relatively high rotational speeds to realize a wider speed range, if the required speed range is not achievable with conventional field weakening methods. The active length of a radial flux PM brushless machine can be reduced by axially displacing its stator or rotor. Since the PM rotor does not have any electrical connections, the axial displacement of the rotor is easier to realize. This mechanical field weakening method is proposed in several patents [3], [4] and qualitatively evaluated in some publications as in [5]. However, detailed analysis about its functionality is lacking.

According to [5], the mechanical field weakening is an effective way of extending the speed range of PM brushless machines, but its main drawbacks are the additional mass and the additional construction space required for the axial stator/rotor displacement. The integration of a mechanical displacement mechanism is an obvious drawback due to its additional mass and need for a robust mechanical design. However, the required additional

1

construction space can be minimized by a proper design, if the available construction space in the drive system is utilized for implementing the displacement mechanism and also for the axial rotor displacement. For example, in an interior rotor PM brushless machine, the displacement mechanism can be built between shaft and rotor core, and the construction space under the end winding conductors can be used for the axial rotor displacement. The most important benefit of applying mechanical field weakening is that the speed range of a PM brushless machine is extended without changing its design in the constant torque operating range. Accordingly, an electric machine with a high torque/power density and a wide speed range is realized. Moreover, mechanical field weakening can be applied to enhance the fault tolerance capability of the electric machine. For example, in case of a winding short-circuit, it is possible to limit the short-circuit current by shortening the active machine length mechanically. Additionally, apart from the displacement mechanism, no additional components and machine parts are needed, and the conventional control algorithms can be applied in the complete operating range.

The analysis of PM brushless machines with an axially displaceable stator/rotor is complicated because of the three dimensional (3-D) magnetic field distribution in the electric machine due to the axial misalignment of stator and rotor. As a result, the flux density distribution in the machine parts is axially inhomogeneous, and the PM flux is not only confined to the stator and rotor cores. The 3-D magnetic field distribution affects both the achievable speed range with a limited construction space and the system efficiency. Therefore, the practical usage of the mechanical field weakening method and its advantages over conventional field weakening methods can only be evaluated after a detailed analysis considering its functional limits and its efficiency.

The objective of this study is to analyze the effectiveness and the limitations of the mechanical field weakening method. In order to achieve this, the effects of the axial rotor displacement on the machine parameters, operational characteristics, speed range and efficiency of a prototype brushless DC (BLDC) machine with an axially displaceable rotor (ADR) [6] are examined in detail. Finally, in the light of these results, possible design optimizations are proposed.

Outline

In chapter 2, requirements of electrical drives for vehicle traction applications and the reasons for the application of PM brushless machines in electrified powertrains are discussed. Then, a literature survey on conventional and alternative field weakening methods is presented. Finally, the basic operating principles and the speed limits of the analyzed mechanical field weakening method are explained.

In chapter 3, the prototype ADR-BLDC machine drive is introduced. The mechanical design of the prototype electric machine and the configuration of the power electronic components are presented. Furthermore, the basic equations of ADR-BLDC machine drives are derived.

In chapter 4, the developed 2-D and 3-D FEM analysis models are introduced. Then the effects of the axial rotor displacement on the magnetic field distribution are examined by using these models.

The electrical parameters of the ADR-BLDC machine are studied in chapter 5. Back EMF waveforms at varying axial rotor positions are calculated by 3-D FEM analysis. Then their dependency on the axial rotor position and affecting geometrical factors are examined in detail. Furthermore, the calculated back EMF waveforms are used to determine the torque constant of the ADR-BLDC machine as a function of the axial rotor position. In order to calculate the self- and mutual inductance values, a model reduction method is developed and used to determine the axial position dependency of the phase inductances. Finally, the calculated back EMF waveforms and the determined no-load inductances are validated by experimental results.

In chapter 6, the influence of the axial rotor displacement on the operational characteristics and the speed limits of the ADR-BLDC machine with different field weakening strategies is examined. A developed dynamic simulation model, which includes the calculated electrical machine parameters from the previous chapter, is used for this analysis. Finally, the simulated phase current characteristics are validated by measurement results.

In chapter 7, the dependency of the losses in the ADR-BLDC machine on the axial rotor displacement is analyzed. Main attention is given to the additional loss mechanisms in the mechanical field weakening range and to the numerical and experimental methods that are developed to determine them. First, the losses with axially aligned stator and rotor, and then the losses in the mechanical field weakening range are identified. Finally, the measurement results are used to validate and to extend the numerical results.

Design recommendations for ADR PM brushless machines are presented in chapter 8. The loss analyses show that the additional losses during mechanical field weakening are critical, unless certain design measures are taken. Therefore, possible design changes to reduce the additional losses are proposed. Then, the speed limits of the electrical, mechanical and combined mechanical field weakening methods are compared, and the field of application of mechanical field weakening is discussed. Finally, additional design considerations for ADR PM brushless machines are highlighted.

Detailed information on the used measuring instruments are provided in the appendix.

2 Electric Machines in Vehicle Traction Applications

In this chapter, the required characteristics of electrical drives for vehicle traction applications are introduced, and pros and cons of PM brushless machines in this field of application are discussed. Moreover, a comprehensive literature survey on the methods that are used to extend the speed range of PM brushless machines including both the conventional and alternative field weakening methods is presented.

2.1 Required Characteristics

The characteristics of electrical drive systems required for vehicle traction applications are studied in many publications as in [7], [8] and [9]. Accordingly, one of the important design criteria is the torque/speed characteristics. Fig. 2.1 shows typical torque/speed and power/speed characteristic curves of a traction motor. The rated torque curve indicates the maximum continuously deliverable torque, whereas the operating region

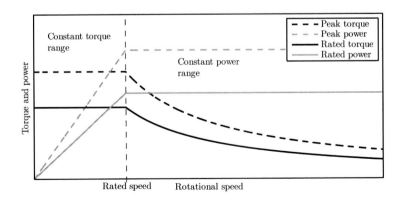

Figure 2.1: Required torque/speed characteristics for vehicle traction applications

between the rated and maximum torque curves is attainable for a pre-defined duration that represents the overload capability of the drive system. Maximum power of the drive system increases proportional to rotational speed and reaches its rated value at rated speed. The operating region, in which the drive system can deliver maximum torque, is called constant torque range. Since power output reaches its maximum at rated speed, characteristic torque curves decay inversely proportional to rotational speed beyond this operating point. This operating region is called constant power range. Considering these operational characteristics, electrical drives need to have:

- high torque at starting and at low rotational speeds in the constant torque range,
- high power at high speeds for cruising,
- wide speed range with a constant power operating range of around 3 to 4 times the base speed range,
- intermitted overload capability.

In literature, base speed and field weakening ranges are usually used synonymously for constant torque and constant power ranges, respectively. Nevertheless, in this study these terms are used differently. Constant torque range and constant power range are used to define the operating regions that are defined according to torque/power specifications of the drive system, whereas base speed range and field weakening range are defined according to the applied control strategy. Maximum achievable rotational speed without field weakening is called base speed; and the operating region up to the base speed is called base speed range. Span of this region depends not only on the design of the electric machine and the inverter, but also on the drive system configuration, the supply voltage and the applied control strategy.

The above defined torque/speed characteristic can be realized by a single-speed transmission, which is preferable because of the lower powertrain weight, volume and cost. On the other hand, a multi-speed gearbox can be used to enhance the performance of the powertrain as well as to optimize the distribution of the operating points in a driving cycle [10].

Efficiency of drive systems in motor and generator modes is demanded to be high over a wide speed and torque range. Instead of optimizing the whole operating range, frequently driven operating regions can be optimized in the design process [11]. These operating regions highly depend on the driving cycle and vehicle properties. As demonstrated in [8], a drive system frequently operates at low speed and moderate torque in a city driving cycle, while it operates at high speed and low torque in a highway driving cycle.

High power and torque densities are among the requirements due to desired low weight of automobiles and a limited construction space. Reasonable and stable production cost has very high importance for the series production. Additionally, electric drives need to fulfill high security and comfort requirements, such as high reliability and robustness at vehicle operating conditions, low torque ripple and low acoustic noise emission.

2.2 PM Brushless Machines

The choice of the appropriate electric drive topology for vehicle traction applications is analyzed in many studies. In [7], [12] and [13], PM brushless machines are reported to have higher efficiencies and torque densities at full load compared to other electric machine types. Because of these characteristics, PM brushless machines are predominantly implemented in conventional EVs, HEVs and PHEVs, as reported in [2], [1] and [14]. Some conventional vehicles with electrified powertrain have induction machine (IM) drives [15], [16] and externally excited synchronous machine (SM) drives [17], [18] and [19]. The main advantage of these electric machine types over PM brushless machines is the absence of costly permanent magnet material.

PM brushless machines inherently have high torque density and efficiency because of the absence of a rotor winding and high energy density PM materials. On the other hand, high cost of magnet materials, their limited overload capability and their limited extended speed range with a limited DC-link voltage and a limited drive system current are the design challenges. There have been many researches to extend the speed range of these machines without affecting their benefits. A detailed literature survey on these studies is included later in this section.

2.2.1 Classification

Based on the direction of the main air-gap magnetic flux, PM brushless machines are categorized as radial flux machines (RFMs) and axial flux machines (AFMs). RFMs are the most common type of the PM brushless machines. On the other hand, AFMs are favorable when the axial length of the machine is limited as well as in high frequency and low speed applications owing to their better torque density [20].

PM brushless machines are mainly classified into two groups: brushless DC machines (BLDCs) and brushless AC machines (BLACs), also known as PM synchronous machines (PMSMs). The main difference between these two machine types is the waveform of the back EMF. BLDC machines are designed to ideally have trapezoidal back EMF and are supplied with rectangular phase currents, whereas BLAC machines ideally have sinusoidal back EMF and are supplied with sinusoidal phase currents for the optimum torque generation. These ideal back EMF waveforms are approximated by both winding configuration and rotor design. For example, a full span stator winding configuration and a surface mounted PM rotor can be implemented to achieve a more or less trapezoidal back EMF waveform. More sinusoidal waveforms can be realized by reducing the winding factor coefficients of higher stator winding harmonics or by reducing spatial air-gap harmonics in the air-gap field.

Another main categorization is made based on rotor topologies into surface mounted PM (SMPM) and interior PM (IPM). Note that these rotor types have various topologies as well. SMPM arrangement is commonly used due to its easier manufacturing and assembly

[21]. IPM rotor PM brushless machines can be designed to have a better field weakening capability than SMPM rotor brushless machines due to their reluctance torque. Moreover, their PMs are shielded from the air-gap magnetic field by the rotor core material; thus, overload capacity of these electric machines is better [7]. Some disadvantages of IPM rotor machines are the higher leakage PM flux in the rotor and the higher torque ripple.

Different stator winding configurations are applied in PM brushless machines. Concentrated non-overlapping configurations with each tooth wound or every second tooth wound are advantageous in reducing the end winding volume. Moreover, they enable better slot fill factors, easier manufacturing and better fault tolerances [22]. On the other hand, lower spatial harmonics in the air-gap field, which result in lower parasitic effects like magnetic noise and additional losses, are achievable with distributed winding configurations [12].

2.2.2 State-of-the-Art Field Weakening Methods

The term "field weakening" is used to describe methods that are applied to increase the maximum rotational speed of an electrical drive above its base speed. According to the required characteristics for traction applications, the field weakening performance of an electrical drive is of high importance.

Electrical Field Weakening

The need for a field weakening operation is explained in the following text by means of a three-phase BLAC machine. First, the basic equations of a three-phase BLAC machine are given, and then the limits of the electrical field weakening operation are discussed with the help of these equations.

The voltage equation of a stator phase in rotor rotating reference frame is given by:

$$u_\mathrm{p} = R_\mathrm{p} i_\mathrm{p} + \mathrm{j}\,\omega_\mathrm{r} L_\mathrm{p} i_\mathrm{p} + L_\mathrm{p} \frac{\mathrm{d} i_\mathrm{p}}{\mathrm{d}t} + e_\mathrm{p} \quad \text{with} \ \ e_\mathrm{p} = \mathrm{j}\,\omega_\mathrm{r} \Psi_\mathrm{PM}, \tag{2.1}$$

where R_p, ω_r, L_p, i_p, Ψ_PM and e_p are the stator phase resistance, pole pair related angular rotor speed, stator phase inductance, stator phase current, PM flux linked by the stator phase winding and induced back EMF, respectively. The pole pair related rotor angular speed, which is equal to the number of pole pairs n_p times the rotor mechanical angular speed ω_m

$$\omega_\mathrm{r} = n_\mathrm{p} \omega_\mathrm{m} \tag{2.2}$$

represents the angular speed that the electrical variables change with, and therefore, it is used in the voltage equation. Eqn. 2.1 can be written in two-dimensional rotor rotating frame (d-q) coordinates as:

$$u_d = R_p i_d - \omega_r L_q i_q + L_d \frac{di_d}{dt}, \tag{2.3}$$

$$u_q = R_p i_q + \omega_r L_d i_d + L_q \frac{di_q}{dt} + e_p. \tag{2.4}$$

The subscripts "d" and "q" stand for direct axis (d-axis) and quadrature axis (q-axis) components, respectively. The electromagnetic torque equation, when amplitude invariant dq-transformation is applied, is given by:

$$T_{em} = \frac{3}{2} n_p \left[\Psi_{PM} - (L_q - L_d) i_d \right] i_q. \tag{2.5}$$

The first part of this equation represents the synchronous torque and the second part represents the reluctance torque. If the rotor is magnetically symmetric, the direct axis inductance L_d and quadrature axis inductance L_q are same. In this case, the reluctance torque equals to zero. In case of a magnetic saliency in rotor ($L_d < L_q$), a positive reluctance torque is produced, if the product of d-axis and q-axis current is negative. Since a positive i_q is required to produce a positive synchronous torque, a negative i_d current is needed to produce a positive reluctance torque.

The operational characteristic of a BLAC machine is mainly limited by two factors: the available DC-link voltage U_{DC} and the current rating of the drive. The absolute stator voltage of a wye-connected BLAC machine is limited by the DC-link voltage as:

$$u_d^2 + u_q^2 \leq \left(\frac{U_{DC}}{\sqrt{3}} \right)^2. \tag{2.6}$$

This equation can be written as

$$(L_d i_d + \Psi_{PM})^2 + (L_q i_q)^2 \leq \left(\frac{1}{\omega_r} \frac{U_{DC}}{\sqrt{3}} \right)^2 \tag{2.7}$$

by using Eqn. 2.1, Eqn. 2.3 and Eqn. 2.4 and neglecting the resistive voltage drop. Eqn. 2.7 shows that the maximum q-axis current i_q can only be supplied up to a certain rotational speed, which is called base speed, unless a negative d-axis current i_d is used. Without field weakening, the q-axis current, and as a result the deliverable synchronous torque, rapidly decrease beyond this rotational speed. So, the main idea of the electrical field weakening is to inject a negative i_d, in order to the extend the rotational speed range with a limited DC-link voltage. The rotational speed range can be extended, until the stator phase current reaches its maximum. The continuous stator phase current is limited by the rated drive current, and this can be shown as:

$$i_d^2 + i_q^2 \leq i_{ra}^2. \tag{2.8}$$

Moreover, the short-time value of the phase current is limited by the peak current of the electrical drive i_{peak}. Due to the current limitation, i_d cannot be increased arbitrarily. To sum up, the maximum achievable rotational speed of a BLAC machine drive system depends on the available DC-link voltage, current rating and machine parameters, especially Ψ_{PM} and L_d.

The d-q transformation and the field oriented control are not applicable to BLDC machines due to non-sinusoidal back EMF waveforms and stator MMF distributions. Instead, the phase advance method is used to extend the rotational speed range beyond the base speed. In this field weakening strategy, the phase currents are advanced with respect to the back EMFs, so that they can reach the desired value during the intervals where the back EMFs are lower than the applied DC voltage. If the back EMF of a phase becomes higher than the applied phase DC voltage, the phase current keeps circulating through the freewheeling diodes due to the stored energy in the phase leakage inductance [23]. Therefore, the value of the phase leakage inductance is important in this field weakening method. The major drawbacks of this method are: the unknown advance angle, the current waveforms significantly deviating from ideal, the decrease in available torque, and the increase in torque pulsations [24], [23].

If the electrical field weakening is applied, no additional components are required. However, the field weakening current reduces the efficiency of the electrical drive at high rotational speeds, and therefore an inverter with a higher current capability might be needed to apply this method. The electrical field weakening method is the mostly applied method that is used to extend the rotational speed range of PM brushless machines, and there are numerous publications, such as [25] and [26], on the optimal design and control of PM brushless machine drives with a wide constant power operating range.

Alternative Field Weakening Methods

Alternative field weakening methods are developed to enhance the PM machine characteristics in the field weakening range without loosing their high efficiency and their high torque/power density. [27], [28], [5] and [29] are good examples in literature, where an overview of alternative methods is presented. These methods are usually realized with a novel electrical drive design, and they can be classified into electrical and mechanical field weakening methods. The alternative field weakening methods for radial flux PM brushless machines are briefly introduced in the following text.

To begin with electrical field weakening methods, the available DC-link voltage can be boosted by a DC/DC converter to extend the rotational speed range, as explained in [30] and [14]. The advantages are reported to be the minimized stress on the inverter due to an additional inverter stage, better efficiency of the electric machine, and the flexible system design. On the other hand, the additional construction space, cost and losses in the DC/DC converter are the drawbacks of this application [31].

Demagnetizing the PMs during operation is another alternative field weakening method. This method is applied in PM memory machines. The PM magnetization is varied by short current pulses either by using the existing AC winding [32] or an additional DC excitation winding [33]. Due to the limited excitation current, PMs with relatively low coercivity, such as aluminum-nickel-cobalt (Al-Ni-Co) and ferrite magnets, are used. Lower energy density of these PMs compared to high energy density PMs, as neodymium-iron-boron (Nd-Fe-B), limits the power/torque density of PM memory machines. In order to overcome this drawback, a PM dual-memory machine is introduced in [34]. This machine is a combination of a PM memory machine and a hybrid machine, so that each rotor pole is composed of Al-Ni-Co and Nd-Fe-B magnets. Magnetization of the Al-Ni-Co magnets is varied by using the additional magnetization winding for variable flux operation, as Nd-Fe-B magnets are implemented to achieve a high power/torque density.

An additional winding is implemented in hybrid excitation synchronous machines (HESMs), in order to weaken or boost the air-gap magnetic flux by adjusting the magnitude and direction of the excitation current. There are various types of these machines, and the common types of them are reported in [35], [36] and [29].

One of the HESM topologies is the synchronous PM hybrid machine, which is a combination of a PM brushless machine and a wound field synchronous machine. These machines are analyzed in [37], [38] and [39]. The main disadvantage of this topology is the need for slip rings and brushes.

Doubly salient PM (DSPM) machines, which are variable reluctance machines with PMs implemented either in the stator or rotor, can be designed as HESM. In [40], two DSPM machine designs with additional DC field excitation in their stators are proposed. Additional DC excitation in the stator is similarly used in [41] to weaken the air-gap magnetic flux as well as to boost it. Additionally, different doubly excited DSPM machine designs with additional DC field excitation are presented in [42], [43] and [44].

Another HESM topology with doubly salient structure and stator PMs is the flux switching PM (FSPM) machine, as reported in [45] and [46]. FSPMs have different PM and stator pole configurations compared to DSPMs. These machine topologies are compared in [47], which reports that FSPMs have better torque density and are more suitable for a AC drive due to their sinusoidal back EMF.

Another HESM with extended field weakening capability is the consequent-pole PM (CPPM) machine. In [48] and [49], a radial flux CPPM machine with two rotor sections and an additional DC excitation winding that is built in the middle of the stator is examined. Each rotor section has PMs with the same orientation and laminated iron poles, alternatively. The relative angular orientation of the rotor sections are arranged, so that the iron poles stand next to the PM poles. The flux created by the DC excitation flows over the iron poles due to the lower magnetic reluctance, and the air-gap flux can be reduced or boosted by changing the direction of the DC excitation current. The drawbacks of CPPM machines are reported to be the reduction in power density due to additional weight and volume of the excitation winding, additional losses due to three-dimensional flux distribution and problems in manufacturing [49].

In addition to the alternative electrical field weakening methods, there are several mechanical field weakening methods in literature. These methods are applied by displacing the passive machine parts to adjust the PM flux paths, by changing the phase terminal connections or by displacing the active machine parts to decrease the stator phase flux linkage. Mechanical methods can enable a large extended speed range especially in high power/torque density PM brushless machines due to the fact that a high field weakening current is not needed to produce a magnetic flux, which opposes the PM flux, or to demagnetize/magnetize the PMs. It is important to note that the methods applied to change the leakage flux paths might have a limited flux weakening capability depending on the machine design. The mechanical displacement of machine parts is usually applied with additional actuators. The main drawbacks of the mechanical methods are the required additional space, weight and cost of the actuators as well as the complexity of mechanical displacement mechanisms.

Inserting flux-shorting iron pieces in the flux barriers is suggested in [50] in order to adjust the air-gap magnetic flux. Similarly, flux-shorting iron end plates are implemented at both sides of the rotor in [51], and the axial position of the end plates is controlled to reduce the air-gap magnetic flux. Another similar approach is applied for switched flux PM (SFPM) machines in [52]. In this study, speed ranges of SFPM machines are extended by mechanically movable flux adjusters, which are the iron pieces outside the stator that are angularly displaced to short a part of the PM flux. Zhu et al. in [53] examine different SFPM topologies in terms of their field weakening capabilities with mechanical flux adjusters. In [54], it is proposed to change the leakage reluctance of the rotor by automatically displacing the magnetic pieces implemented in flux barriers by centrifugal force.

Changing the connections between the phase coil groups and changing the stator phase interconnections is proposed to extend the rotational speed range of PM brushless machines in several publications. In [55], it is proposed to change the connections between coil groups in order to adjust the back EMF of three-phase PM brushless machines. Similarly, the active number of the stator coil turns is reduced to limit the back EMF at high rotational speeds in [56]. The phase interconnection configuration of a three-phase PM brushless machine is changed from wye to delta in [57] by using semiconductor devices in order to increase the applied DC voltage over a stator phase at high rotational speeds, which is called winding changeover or winding change technique. In [58], winding changeover is applied to a five-phase PM brushless machine by mechanical switches by using the advantage of the higher number of interconnection possibilities of the stator phases. The stator winding configuration can be changed during operation either to adjust the back EMF or the applied DC voltage over a stator phase. However, torque interruptions can occur due to variation of the stator phase currents during the switching operation.

Rotating cylindrical PMs during operation is proposed in [5] and [28], where an integrated magnet adjustable bar (IMAB) machine is introduced. Each rotor pole of this machine is composed of two fixed rectangular PMs and a rotatable cylindrical PM that is inserted in the pole middle. The magnetic flux of the fixed PMs is strengthened by arranging the rotatable PMs in the same magnetic orientation as the fixed ones and weakened by rotating the cylindrical magnets.

Changing the relative angular orientation of the rotor sections of split rotor electric machines is suggested in [59] and [60]. The aim of this method is to reduce the magnetic flux that is linked by phases. In [59], the stable relative angular positions of the rotor sections are discussed, and a spring mechanism is implemented to avoid the unwanted relative angular orientation of the rotor sections. Similarly, varying the relative angular positions of the rotor layers, namely inner and outer rotor, is proposed to reduce the phase flux linkage in a patent application that is addressed in [29].

Another patent application included in [29] suggests to form leakage paths between the rotor sections of split rotor electric machines. To realize this, an additional angularly displaceable reluctance type rotor section is built axially in the middle of the rotor sections with PMs.

Three mechanical field weakening methods for DSPM machines with stator PMs are introduced in [61] and [40]. These methods are applied by increasing the PM leakage flux with the help of movable magnetic flux shorting pieces, diverting the PM flux away from the stator with a rotatable magnetic/non-magnetic shell driven by an actuator, and by decreasing the PM flux in the stator by axially sliding the magnets out from the stator.

Reducing the active axial length of PM brushless machines by an axial rotor or stator displacement is proposed in several patents, such as in [3], [4] and [62], and its application in an electric vehicle is reported in [63] and [64]. According to the qualitative evaluations in [5], this mechanical field weakening method has the following advantages:

- It is an effective method to extend the rotational speed range of a PM brushless machine.

- If mechanical field weakening is applied, the inverter power capacity can be reduced.

- The back EMF can be reduced independent of the inverter in case of an inverter fault.

- The short circuit current can be reduced in case of a short circuit in the winding.

Additionally, it can be applied to conventional PM brushless machine drives by implementing an additional axial displacement mechanism and new control strategies are not needed. These are clear advantages of this field weakening method compared to alternative methods with sophisticated designs. On the other hand, the complexity of the displacement mechanism owing to the required high displacement force and the additional construction space needed for the axial rotor displacement are stated as drawbacks of this method in [5]. Consequently, the required additional space for the axial rotor/stator displacement and the mechanical displacement mechanism reduce the torque/power density of the drive system, which conflicts with the high torque/power density feature of PM brushless machines. Therefore, the effectiveness of this method is an important design question.

Due to the fact that a detailed analysis on this promising mechanical field weakening method does not exist, a prototype ADR-BLDC machine with two rotor sections is examined in this study. This analysis focuses on changes in the magnetic properties of the electric machine depending on the axial rotor position with the aim of determining the effects of this field weakening method on the operational characteristics of PM brushless machines.

2.2.3 PM Brushless Machines with Axially Displaceable Stator/Rotor

The basic operation principle of the analyzed mechanical field weakening method has similarities with the field weakening of separately excited DC commutator machines. The rotor PM flux that is seen by the phases is controlled independent of the stator phase currents by adjusting the active axial length of the electric machine by axially displacing its rotor or stator. As the active axial length is reduced, the flux linkage of the stator phases decreases. Consequently, the induced back EMF can be reduced at high rotational speeds in order to extend the speed range with a limited DC-link voltage. Fig. 2.2 shows two rotor displacements of an axially displaceable rotor BLDC (ADR-BLDC) machine with two rotor sections. In the base speed range, the rotor sections are axially aligned with the stator, and this rotor position is called rotor base position. At high rotational speeds, the rotor sections are axially displaced away from each other. The distance between the rotor sections is adjusted depending on the required field weakening level. Exemplarily induced back EMF waveforms at a given rotational speed at rotor base position and with axially displaced rotor are shown in Fig. 2.3.

The mechanical field weakening range is basically limited due to

- the limited phase current and the limited DC-link voltage,

- the limited construction space available for the axial rotor displacement.

The first factor that limits the mechanical field weakening range is the limited phase current and the limited DC-link voltage. The maximum rotational speed, at which the peak power is deliverable, is limited due to the fact that the phase current and the phase voltage reach their maximum before the theoretical zero induced voltage operation point is reached. At this operation point in the mechanical field weakening range, the torque produced at peak power can be written for a 3-phase PM brushless machine with $L_d = L_q$ as:

$$T = \frac{P_{\text{peak}}}{\omega_{\text{m}}} = \frac{P_{\text{peak}} n_{\text{p}}}{\omega_{\text{r}}} = \frac{3}{2} n_{\text{p}} \Psi_{\text{PM}} i_{\text{peak}} \text{ for } i_{\text{d}} = 0\,\text{A}. \tag{2.9}$$

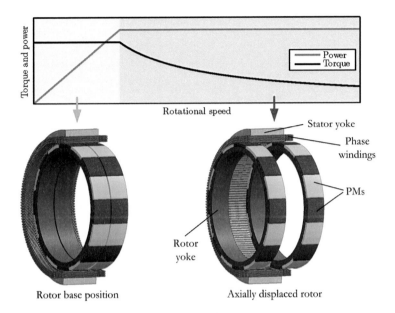

Figure 2.2: Axial rotor positions

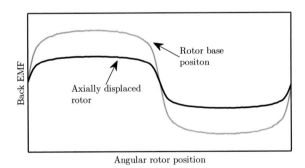

Figure 2.3: Induced back EMF waveforms at a given rotational speed

Consequently, the maximum power can only be delivered at a higher rotational speed, if Ψ_{PM} is reduced inversely proportional to rotational speed. This means that the amplitude of the back EMF must be kept constant. On the other hand, the voltage drop on the q-axis inductance increases with the rotational speed. This leads to a higher phase voltage, thus the speed cannot be further increased. This operation point is illustrated in Fig. 2.4, where

$$\omega_{r1} < \omega_{r2}, \quad \frac{\Psi_{PM}(\omega_{r1})}{\Psi_{PM}(\omega_{r2})} = \frac{\omega_{r2}}{\omega_{r1}}, \quad e_p(\omega_{r1}) = e_p(\omega_{r2}) \quad \text{and} \quad u_p(\omega_{r1}) < u_p(\omega_{r2}). \quad (2.10)$$

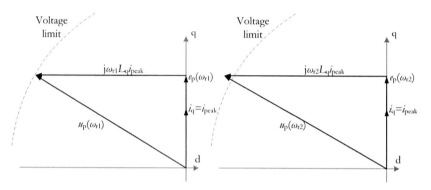

Figure 2.4: Maximum speed at P_{peak} with mechanical field weakening without construction space limitations

The maximum rotational speed, at which the maximum power is deliverable, is calculated neglecting the voltage drop on the phase resistance and the losses in the electric machine for a PM brushless machine with $L_d = L_q$, as follows.

If the d-axis current is zero, Eqn. 2.7 can be written as:

$$(\Psi_{PM})^2 + (L_q i_q)^2 \leq \left(\frac{1}{\omega_r} \frac{U_{DC}}{\sqrt{3}} \right)^2 . \quad (2.11)$$

In this equation, Ψ_{PM} is an unknown variable that can be reduced arbitrarily by reducing the active machine length, provided that there are no construction space limitations. Therefore, Ψ_{PM} is written as a function of the peak power P_{peak}, the rotor angular speed ω_m and the q-axis phase current according to Eqn. 2.5 as:

$$\Psi_{PM} = \frac{T}{\frac{3}{2}n_p i_q} = \frac{P_{peak}/\omega_m}{\frac{3}{2}n_p i_q} = \frac{P_{peak}/\omega_r}{\frac{3}{2}i_q} . \quad (2.12)$$

By using Eqn. 2.11 and Eqn. 2.12, the maximum angular speed at P_{peak} is found as:

$$\omega_{r,max\_at\_P_{peak}} = \sqrt{\frac{\left(U_{DC}/\sqrt{3} \right)^2 - \left(P_{peak}/(\frac{3}{2}i_q) \right)^2}{(L_q i_q)^2}} . \quad (2.13)$$

As a result, $\omega_{r,max\_at\_P_{peak}}$ with an unlimited axial displacement range depends on the

DC-link voltage, the q-axis inductance, and the phase current; and a low q-axis inductance is advantageous for a wide extended speed range. Furthermore, based on the fact that the torque is proportional to the product of Ψ_{PM} and i_q, the minimum Ψ_{PM} occurs when the phase current is maximum. Therefore, the maximum rotational speed at P_{peak} for the minimum plausible Ψ_{PM} can be calculated by using

$$\omega_{\text{r,max\_at\_P}_{\text{peak}}} \text{ for minimum } \Psi_{PM} = \sqrt{\frac{\left(U_{DC}/\sqrt{3}\right)^2 - \left(P_{peak}/(\tfrac{3}{2}i_{peak})\right)^2}{\left(L_q i_{peak}\right)^2}}. \tag{2.14}$$

The equations above can be similarly written for the rated power case. Beyond this operation point, the rotational speed can only be increased, if the power output is reduced. The maximum speed at P_{peak} and the third operating range, where the power output is reduced, are illustrated in Fig. 2.5.

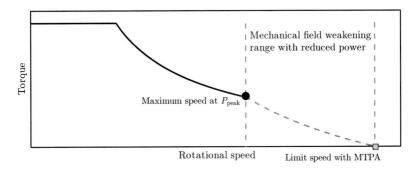

Figure 2.5: Torque speed characteristics with mechanical field weakening without construction space limitations

The limit rotational speed with mechanical field weakening is reached in the reduced power field weakening range, if the produced torque is equal to zero. This speed is theoretically infinite with an unlimited axial rotor displacement range neglecting the machine losses, since both Ψ_{PM} and i_q can be reduced to zero. Due to the losses in the electric machine, this region is limited in practice. Moreover, operating points at high speeds and low power output values are not interesting for traction applications. A relevant speed limit for traction applications at $T{=}0$ Nm can be calculated, if the maximum torque per ampere (MTPA) control strategy is applied. In this case, the value of the Ψ_{PM} at $\omega_{\text{r,max\_at\_P}_{\text{peak}}}$ is kept constant, and phase currents are reduced to achieve higher rotational speeds than $\omega_{\text{r,max\_at\_P}_{\text{peak}}}$. The maximum speed at $T{=}0$ Nm with MTPA can be determined by using

Eqn. 2.11 as:

$$\omega_{\text{r,lim\_MTPA}} = \frac{U_{\text{DC}}}{\sqrt{3}\,\Psi_{\text{PM}}(\omega_{\text{r,max\_at\_P}_{\text{peak}}})} \qquad\qquad (2.15)$$

This limit speed with MTPA is also marked in Fig. 2.5, where the dashed line represents the maximum torque/speed curve achievable, if Ψ_{PM} is equal to its value at $\omega_{\text{r,max\_at\_P}_{\text{peak}}}$.

Furthermore, the limits of the mechanical field weakening must be corrected, if the minimum phase flux linkage is not achievable within the range of the axial rotor displacement. Actually, it is preferable to limit the axial rotor displacement in order to ensure a higher volumetric torque/power density, lower additional losses and a simpler rotor displacement mechanism.

Axially displaceable stator/rotor PM brushless machines can be designed in several ways. First of all, the stator or the rotor can be axially displaced, but axial rotor displacement is more common due to lacking electrical connections of the rotor. Additionally, in [62] various design possibilities, such as interior and exterior rotor topologies, designs with one part and split PM rotors, and designs with various air-gap forms are presented.

An important design feature of axially displaceable stator/rotor PM brushless machines is the mechanical displacement mechanism, which is commonly applied with the help of additional actuators as in [3].

3 Prototype Electrical Drive System

In this study, an existing prototype electrical drive system is used for the performance analysis of axially displaceable rotor BLDC (ADR-BLDC) machines. The drive system consists of a multi-phase ADR-BLDC machine, power electronic components, sensing devices, a single speed gearbox, and a cooling system. This prototype is developed by the Group Research of Volkswagen AG within the scope of a project funded by the Federal Ministry for the Environment, Nature Conservation and Nuclear Safety of the Federal Republic of Germany. The design objectives and specifications, the electromagnetic and mechanical design, the production of the prototypes, and the elementary test results of the electrical drive are reported in the final project report [6] in detail.

In this chapter, the design specifications, design, and materials of the ADR-BLDC machine as well as the configuration of the power electronic components are introduced. Additionally, basic mathematical equations of ADR-BLDC machine drives are derived.

3.1 Design Specifications

The design specifications of the drive system are listed below.

Construction space specifications:

- Radius = 280 mm
- Axial length = 110 mm

Output specifications:

- Rated power = 50 kW
- Peak power = 85 kW
- Rated torque = 150 Nm
- Peak torque = 300 Nm
- Maximum rotational speed = 12000 rpm

DC supply specifications:

- Peak input DC current = 500 A

- Nominal DC voltage $= 300\,\mathrm{V}$
- Peak DC voltage $= 350\,\mathrm{V}$

3.2 ADR-BLDC Machine: Design and Materials

The analyzed prototype is a 10-phase, 8-pole pair ADR-BLDC machine with an SMPM rotor. The main objective when designing the ADR-BLDC machine was to build a high torque/power density electric machine for automobile traction applications. In order to achieve this, a high speed, multi-phase and multi-pole BLDC machine with an axially displaceable rotor is preferred. The reasons for these choices are as follows:

- A high speed design is chosen since the size of the electric machine for a given power output rating reduces with increasing rated rotational speed due to the reduction in the torque rating.

- It is known that BLDC machines have a higher power/torque density than BLAC machines. This can be explained with the fact that the rms value of the air-gap flux density of a BLDC machine is higher given that both machines have the same peak air-gap flux density [23]. In order to compare the power/torque density of BLDC and BLAC machines, the power output ratio of these machines is derived by equating the copper losses in their stators, where it is assumed that the electric machines have ideal back EMF and current characteristics, the same phase resistance, and a unity power factor, as follows.

  Since the copper losses are assumed to be equal, the rms values of the phase currents are also equal as given by

$$I_{\mathrm{BLDC_{rms}}} = I_{\mathrm{BLAC_{rms}}} \,. \tag{3.1}$$

  If the BLDC machine is supplied with 120° current pulses and the BLAC machine with sinusoidal phase currents, Eqn. 3.1 can be written in terms of the peak current values as follows:

$$\hat{I}_{\mathrm{BLDC}}\sqrt{\frac{2}{3}} = \hat{I}_{\mathrm{BLAC}}\sqrt{\frac{1}{2}} \tag{3.2}$$

$$\frac{\hat{I}_{\mathrm{BLDC}}}{\hat{I}_{\mathrm{BLAC}}} = \frac{\sqrt{3}}{2} \tag{3.3}$$

  Then, the output power ratio is

$$\frac{P_{\mathrm{BLDC}}}{P_{\mathrm{BLAC}}} = \frac{2 \times \hat{e}_{\mathrm{p}} \times \hat{I}_{\mathrm{BLDC}}}{3 \times \hat{e}_{\mathrm{p}}/\sqrt{2} \times \hat{I}_{\mathrm{BLAC}}/\sqrt{2}} = \frac{2}{\sqrt{3}} = 1.1547 \,. \tag{3.4}$$

Accordingly, BLDC machines have almost 15.4% higher power than BLAC machines in this case. Similarly, the output torque ratio is higher in BLDC machines.

- The torque/power density of BLDC machines can be further increased by increasing the number of phases, as discussed in [65], [66] and [24]. This increase in torque/power density is mainly achieved due to the following reasons:

    1. By increasing the number of phases, the actively used copper volume is increased at any time instant. For example, only 2 out of 3 phases of a 3-phase machine are conducting at any time, while this number is 4 out of 5 phases for 5-phase machines. This is the case in an electric machine with wye-connected phase terminals without a neutral connection, due to the fact that the phase currents must add up to zero. It is important to note that this discussion is also true for the even phase numbers, i.e. the system is already reduced. A system is reduced, i.e. there are no phases that are spatially shifted by 180° with respect to each other, as explained in [67].

    2. Time harmonics of the phase current with an order smaller than the phase number produce forward going synchronous MMF space harmonics. Therefore, they produce synchronous torque. Higher order time harmonics similarly produce a synchronous torque, but at the same time they result in an asynchronous torque with a higher amplitude, causing additional losses, torque ripples, and noise emissions [65]. Accordingly, the 3rd current harmonic in a 5-phase machine and the 3rd and 5th current harmonics in a 7-phase machine produce a synchronous torque with an appropriate machine design and control.

- With increasing pole number, the height of the stator yoke can be reduced due to the partition of the total flux among the pole pairs, which results in shorter flux paths and lower flux density in the stator yoke.

- A wider speed range can be achieved by applying mechanical field weakening without changing the design of electric machines in the base speed range. As a result, a compact design with a high extended speed range is possible. At this point, the additionally required construction space for the axial rotor displacement must be considered. Therefore, the space below the end winding conductors is utilized for this purpose in the prototype ADR-BLDC machine.

Additionally, the electric machine and the power electronic components are integrated, and a common cooling system is used to achieve a compact design, as presented in [68].

Fig. 3.1 and Fig. 3.2 show the mechanical arrangement of the ADR-BLDC machine. It is an interior rotor electric machine. The stator has in total 160 teeth, and the slot openings have a constant width, so that wires with rectangular cross-section fit into the slots. The SMPM rotor is split into two symmetrical parts. This design enables an optimal usage of the construction space below the end winding conductors at both stator sides. A BLDC machine design with an SMPM rotor is chosen, since a high torque/power density can be achieved with SMPM rotor PM brushless machines due to the low PM leakage flux in the rotor. On the other hand, this machine type has a limited extended speed range because of the high flux linkage and relatively low phase inductances, if conventional electrical field weakening methods are applied, as discussed in [69] and [55]. Therefore, it is reasonable to apply the mechanical field weakening method to an SMPM rotor machine.

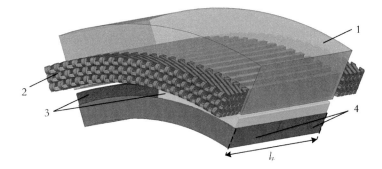

Figure 3.1: Mechanical arrangement of the ADR-BLDC machine at rotor base position: 1) stator core, 2) stator winding, 3) PMs, 4) rotor yoke, and l_z is the axial length of stator core and rotor

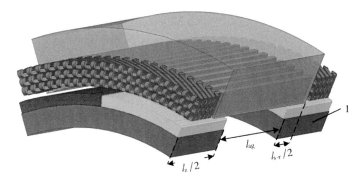

Figure 3.2: Mechanical arrangement of the ADR-BLDC machine with an axially displaced rotor: 1) rotor overhang, l_{ag} is the axial gap between the rotor parts, and l_{s-r} is the axial length of the stator/rotor overlapping

The ADR-BLDC machine has a full pitch overlapping wave winding. The unsymmetrical and symmetrical phase vector diagrams of this machine are shown in Fig. 3.3. As can be seen in this figure, the consequent stator phases are circumferentially shifted by an angle

α, which is defined as:

$$\alpha = \frac{\pi}{Nn_\mathrm{p}} = \frac{\pi}{80} \text{ rad} \tag{3.5}$$

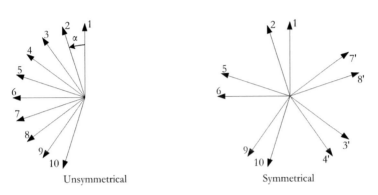

Unsymmetrical   Symmetrical

Figure 3.3: Phase vector diagram of the ADR-BLDC machine

The coils of a phase are placed into $2n_\mathrm{p} = 16$ slots, where only the coils of one phase are located in each slot, as shown in Fig. 3.4. The six-layer winding configuration, which is implemented to achieve a high phase flux linkage, can also be seen in this figure. This winding configuration enables trapezoidally induced voltage waveforms. Moreover, the magnetic couplings between the phases are lower than chorded winding configurations. This simple winding design enables also shorter end winding turns. The length of the end winding conductors is an important design criterion in order to achieve a compact design with high efficiency. Therefore, the end winding conductors must be as short as possible. Indeed non-overlapping winding configurations are known to have shorter end winding conductors. However, an overlapping winding configuration is chosen for the prototype machine due to the lower spatial harmonics in the air-gap field, and appropriate measures to shorten the end winding length are taken. The coils are made of a special Litz wire with a diameter of 0.3 mm because of the high fundamental frequency seen by the stator that is equal to 1600 Hz at maximum rotational speed, low phase inductance values, and non-sinusoidal phase current waveforms. Additionally, the used Litz wire has a rectangular cross-section, which makes it possible to realize slot fill factors over 78%, as presented in [70] and [22].

The axial rotor displacement of the ADR-BLDC machine is achieved hydraulically by using centrifugal force as explained in [6]. In this displacement mechanism, an active actuator for axial rotor positioning is not required. However, the axial rotor position must be monitored, which can be done by using distance sensors or by measuring the voltage induced in a search coil.

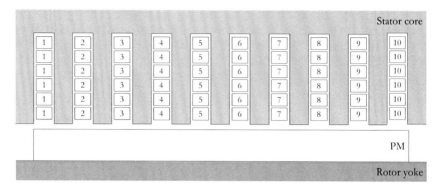

Figure 3.4: Winding configuration over a pole

The mechanically integrated electric machine and power electronic components are cooled by a liquid coolant in the stator water jacket.

In total, 10 digital Hall effect sensors are built into the non-magnetic slot wedges in the slots openings of the ADR-BLDC machine. The temperature of the electric machine is monitored by using 4 temperature sensors that are integrated between the end winding conductors.

The stator is made of Fe-Si alloy 10JNEX900 [71], which is a non-oriented electrical steel with a thickness of 0.1 mm and 6.5% silicon content. Owing to the high silicon content and the low lamination thickness, this material is suitable for high frequencies up to 40 kHz [72]. The rotor yoke is made of C45, which is a medium carbon steel with 0.45% carbon content. An electrical steel is not used in the rotor yoke, since the eddy currents induced in this part are not critical for SMPM machines due to the large effective air-gap length. Fig. 3.5 shows the magnetization curves of electric steel 10JNEX900 and steel C45. These characteristics are used to define the magnetic characteristics of the soft magnetic materials in the FEM analsysis models.

Radially magnetized Nd-Fe-B PMs are installed in the ADR-BLDC machine. Nd-Fe-B PMs are commonly used in PM electric machines due to their high energy density, high remanent flux density B_r and high coercivity H_c. The major drawback of these PMs is their limited operating temperature, which must be considered in the design in order to avoid demagnetization. For this reason, the installed PMs are segmented in axial direction, in order to reduce the eddy current losses caused by high order spatial harmonics in the air-gap field.

The housing of the ADR-BLDC machine is made of aluminum casting alloy AlSi10Mg. This material is used because of its good mechanical properties, such as good thermal conductivity, low weight and high strength. Beside the mechanical properties, the magnetic

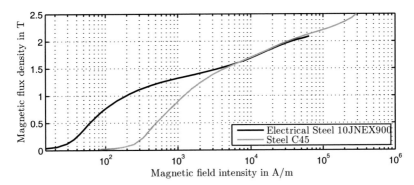

Figure 3.5: DC magnetization curves of the soft magnetic materials: electrical steel 10JNEX900 and steel C45 [72]

and electric properties of the housing material are important, since the housing is exposed to a time-varying magnetic flux in the mechanical field weakening range [73]. According to [74], the electrical conductance of this alloy is between $16-21 \times 10^6$ S/m. The main elements of this alloy are non-ferromagnetic. The ferromagnetic materials in the composition are iron (less than 0.5%) and nickel (less than 0.05%). Due to the low amount of the ferromagnetic materials, the housing material is considered as non-ferromagnetic in this study.

Table 3.1 summarizes the materials of the ADR-BLDC machine and some of the physical properties of these materials.

Part	Material	ρ kg/m^3	σ S/m	μ_{r_max}	λ $W/(m \cdot K)$
Stator core	10JNEX900	7490	1.92×10^6	23000	-
Rotor yoke	Steel C45	7850	5.55×10^6	$3500 - 20000$	-
PM	BM 38SH	7350	$0.63 - 0.71 \times 10^6$	≈ 1.1	$8 - 10$
Housing	AlSi10Mg	2680	$16 - 21 \times 10^6$	≈ 1	$130 - 150$

Table 3.1: Materials of the ADR-BLDC machine

3.3 Power Electronic Components

The power electronic components of the drive system are schematically shown in Fig. 3.6. As can be seen, the stator phases are fed by Insulated-Gate Bipolar Transistors (IGBTs) based full-bridge voltage source inverters. The major disadvantage of this configuration is the number of the semiconductor devices. If each phase is fed by a full-bridge inverter, the number of the power electronic modules and the diodes will be twice as high as in a multi-leg inverter. On the other hand, the phases are electrically isolated, and there are no voltage or current constraints due to the internal connection of the stator terminals. The power electronic modules and their gate drivers are no commercial products but especially developed for the drive system mainly because of the construction space constraints.

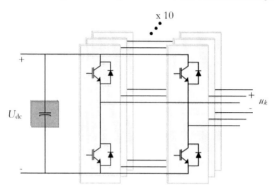

Figure 3.6: Configuration of the power electronic components

A ring shaped polypropylene thin film capacitor with integrated terminal connections is implemented in the drive system. The advantages of this power capacitor are its better mechanical system integration, low internal inductance and short terminal connections [75].

3.4 Basic Equations and Modeling

The known mathematical equations derived for BLDC machines can be used to model the ADR-BLDC machine in the whole operating range. However, the characteristics of the electric machine change during the mechanical field weakening operation, since the machine parameters depend on the axial rotor position. Therefore, the effects of the axial rotor displacement on the characteristics of BLDC machines are mentioned as well. For simplicity reasons, all losses, except for the stator winding copper losses, and the effect of the induced eddy currents are ignored in the derivation of the equations in this section.

BLDC machines are modeled by using phase variables owing to non-sinusoidal flux distribution. The phase voltage equation of phase k is given by

$$u_k = R_\mathrm{p} i_k + \frac{\mathrm{d}\Psi_k}{\mathrm{dt}}, \tag{3.6}$$

where Ψ_k is the total flux linkage of phase k. The flux linkage of a phase is composed of the leakage flux of the stator phase winding Ψ_s and the flux linkage due to the PM flux in the stator Ψ_{PM}, as given by

$$\Psi_k = \Psi_{s,k} + \Psi_{PM,k}\,. \tag{3.7}$$

The stator leakage flux can be written in terms of the phase inductance L_k and the phase current i_k, if only phase k is supplied with current as

$$\Psi_{s,k} = L_k i_k\,. \tag{3.8}$$

By using Eqn. 3.7 and Eqn. 3.8, Eqn. 3.6 can be written as

$$u_k = R_p i_k + L_k \frac{di_k}{dt} + \frac{dL_k}{dt} i_k + \frac{d\Psi_{PM,k}}{dt}\,. \tag{3.9}$$

The last term in this equation is defined as the induced voltage or the back EMF e_k of phase k. The back EMF equation can be written as a function of the rotor angular position θ_r and the rotor rotational speed ω_r as

$$e_k = \frac{d\Psi_{PM,k}}{dt} = \frac{d\Psi_{PM,k}}{d\theta_r}\,\omega_r\,. \tag{3.10}$$

The phase flux linkage due to PM flux alternates periodically with electrical frequency due to the rotation of the PM rotor. Therefore, the induced voltage has the same periodicity. Moreover, its amplitude depends linearly on rotational speed, if the induced eddy current reaction field is neglected.

Eqn. 3.9 is written for all stator phases in Eqn. 3.11, if all phases are supplied with currents, where M_{k-l} is the mutual inductance between phase k and phase l due to magnetic coupling.

$$\begin{bmatrix} u_1 \\ u_2 \\ \vdots \\ u_{10} \end{bmatrix} = R_p \begin{bmatrix} i_1 \\ i_2 \\ \vdots \\ i_{10} \end{bmatrix} + \underbrace{\begin{bmatrix} L_1 & M_{1-2} & \cdots & M_{1-10} \\ M_{2-1} & L_2 & \cdots & M_{2-10} \\ \vdots & \vdots & \ddots & \vdots \\ M_{10-1} & M_{10-2} & \cdots & L_{10} \end{bmatrix}}_{\text{Inductance matrix}} \frac{d}{dt} \begin{bmatrix} i_1 \\ i_2 \\ \vdots \\ i_{10} \end{bmatrix}$$

$$+ \begin{bmatrix} \frac{d}{dt} \begin{bmatrix} L_1 & M_{1-2} & \cdots & M_{1-10} \\ M_{2-1} & L_2 & \cdots & M_{2-10} \\ \vdots & \vdots & \ddots & \vdots \\ M_{10-1} & M_{10-2} & \cdots & L_{10} \end{bmatrix} \end{bmatrix} \begin{bmatrix} i_1 \\ i_2 \\ \vdots \\ i_{10} \end{bmatrix} + \begin{bmatrix} e_1 \\ e_2 \\ \vdots \\ e_{10} \end{bmatrix} \tag{3.11}$$

Since the stator and rotor materials have non-linear magnetic characteristics, the relationship between flux linkage and phase currents is non-linear. Consequently, the inductance of a winding can be defined in two different ways. The first definition is the apparent inductance L_{app} defined by

$$L_{app} = \frac{\Psi}{i} \, . \tag{3.12}$$

The second one is the differential inductance L_{diff} defined by

$$L_{\text{diff}} = \frac{d\Psi}{di} \, . \tag{3.13}$$

The apparent inductance is used to calculate the total flux linkage depending on the current, whereas the differential inductance represents the amount of flux linkage change caused by a change in the related current. According to the second term in Eqn. 3.11, differential inductances are used in equivalent circuit models of electric machines. Consequently, the differential inductances are calculated in this study, and they are denoted without subscripts for simplicity reasons.

Phases of electric machines are magnetically coupled. Consequently, the flux linkage of a phase is caused by the current of this particular phase and by the currents of the other phases. The relationship between the flux linkage of phase k and the current of the same phase is defined by the phase self-inductance L_k as

$$L_k = \frac{d\Psi_k}{di_k} \, , \tag{3.14}$$

and the magnetic coupling between two stator phases k and l is defined by the mutual inductance M_{k-l} as

$$M_{k-l} = \frac{d\Psi_k}{di_l} \text{ and } M_{l-k} = \frac{d\Psi_l}{di_k} \, . \tag{3.15}$$

Owing to the symmetrical design of the stator phases of the ADR-BLDC machine, M_{k-l} and M_{l-k} are equal. If stator winding is of a non-overlapping type, magnetic phase couplings are relatively low. Moreover, if alternate teeth are wound, magnetic couplings can be designed to be negligibly small [76]. On the other hand, the ADR-BLDC machine has 10 stator phases with an overlapping full-pitch wave winding design; thus magnetic phase couplings are needed to be considered.

The ratio between self- and mutual inductances is an important parameter which determines operational characteristics of the electric machine. This ratio is called coupling factor k, and it is defined as

$$k_{k-l}(\theta_{\text{r}}) = \frac{M_{k-l}(\theta_{\text{r}})}{\sqrt{L_k(\theta_{\text{r}})L_l(\theta_{\text{r}})}} \, . \tag{3.16}$$

Given that the phase inductances are constant, the third term in Eqn. 3.11 disappears. However, despite the magnetically symmetrical rotor design, inductance values vary with the angular rotor position due to the saturation of the stator and rotor cores.

The instantaneous back EMF can be written as in Eqn. 3.17 by using the machine constant $k\Phi$ and a rotor position dependent function $f(\theta_r)$ with a maximum magnitude of ± 1. This function represents the rotor position dependency of the back EMF.

$$e = k\Phi\, f(\theta_r)\, \omega_m \tag{3.17}$$

The idealized back EMF and phase current waveforms of the ADR-BLDC machine are shown in Fig. 3.7 over an electrical period. Since the phases are electrically decoupled from each other, there are no phase current constraints due to the interconnection of the phases. Therefore, the current conduction interval can be chosen independently of the phase number.

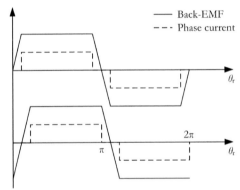

Figure 3.7: Idealized back EMF and phase current waveforms of two adjacent stator phases

Electrical power converted into mechanical power is given by

$$T\omega_m = \sum_{k=1}^{10} e_k i_k \tag{3.18}$$

and the instantaneous value of the torque can be calculated by

$$T = \frac{1}{\omega_m} \sum_{k=1}^{10} e_k i_k = k\Phi \sum_{k=1}^{10} f(\theta_r, k)\, i_k . \tag{3.19}$$

If current commutation is successfully achieved in each stator phase, the produced torque can be calculated by using the machine constant and the DC input current I_{dc} as

$$T = k\Phi\, I_{dc}, \text{ where } I_{dc} = \sum_{k=1}^{10} |i_k| . \tag{3.20}$$

Fig. 3.8 shows ideal DC input current characteristics of ADR-BLDC machines. As can be seen, the supplied DC current must be kept constant in the mechanical field weakening range since the machine constant in Eqn. 3.20 decreases, as the rotor is axially displaced.

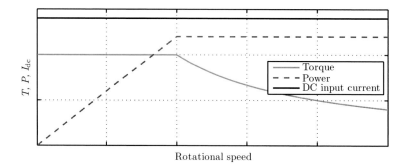

Figure 3.8: Ideal DC current characteristics of an ADR-BLDC machine

Finally, the motion equation of a system with a total moment of inertia J and a load torque T_{L} is given by

$$J \frac{\mathrm{d}\omega_{\mathrm{m}}}{\mathrm{d}t} + T_{\mathrm{L}} = T, \quad \omega_{\mathrm{m}} = \frac{\mathrm{d}\theta_{\mathrm{m}}}{\mathrm{d}t}. \tag{3.21}$$

Only the dependency of the electric variables on the angular rotor position is considered so far. In addition to this, the machine parameters may change with the axial rotor position, and this dependency must be considered during quasi-stationary operation.

The transient behavior of the drive system during the axial displacement of the rotor must also be considered in the modeling. In automobile applications, an acceleration from stand still up to maximum speed takes seconds, hence transient effects are uncritical. This can be shown by the following example, where the influence of the axial rotor displacement on the back EMF is analyzed. For simplicity reasons, the flux linkage waveform is approximated with a triangular wave with an amplitude $\hat{\Psi}_{\mathrm{PM}}$. During an acceleration from 4000 rpm to 12000 rpm with a constant acceleration rate in 5 s, the amplitude of the flux linkage is mechanically decreased to one third $\hat{\Psi}_{\mathrm{PM}}/3$. The induced voltage difference Δe in a stator phase due to axial rotor displacement during this acceleration process can be calculated by

$$\Delta e = \frac{\Delta \Psi_{\mathrm{PM}}}{\Delta t} = \frac{(2\hat{\Psi}_{\mathrm{PM}}/3)}{\Delta t}. \tag{3.22}$$

By inserting an approximate $\hat{\Psi}_{\mathrm{PM}}$ value of 0.1 Vs, the induced voltage difference is around 13 mV, which is negligibly small for a 300 V application. Since the electrical period of the electrical variables is much lower than the axial rotor displacement duration, the influence of the axial rotor displacement on the transient behavior of ADR-BLDC machines can be neglected.

Definition of the Reference Position

The reference position of a phase is defined to be in the pole-gap, as shown in Fig. 3.9. The rotor angular position dependency of the machine parameters is illustrated according to this reference position in this study.

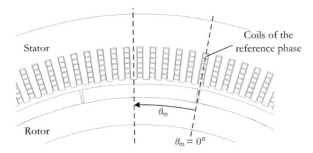

Figure 3.9: Definition of $\theta_m = 0°$ of a phase

The parameters, such as the induced back EMF and the phase inductance as well as the magnetic field related values, change periodically with the pole pair n_p related angular rotor position that is defined by

$$\theta_r = \theta_m n_p, \tag{3.23}$$

also called electrical angular rotor position. Therefore, these variables are illustrated just for an angular span of a pole pair which is equal to $360°/n_p = 45°$. Accordingly, the conductors of the reference stator phase are in the pole-gaps at $\theta_m = 0°$ and $\theta_m = 22.5°$, and in the pole middle at $\theta_m = 11.25°$ and $\theta_m = 33.75°$.

4 Numerical Field Calculation

The analytical electromagnetic analysis of electric machines is either possible for simple field problems or with simplified electric machine geometries and material characteristics. Therefore, numerical field calculations are widely used for calculating electromagnetic fields in electric machines. The most known numerical field calculation methods are the Finite Differential Method (FDM), the Finite Element Method (FEM), the Finite Integral Method (FIM) and the Boundary Element Method (BEM), among which FEM is the most commonly used method for electromagnetic analysis.

In the following chapters, electromagnetic field distributions, parameters and losses of the ADR-BLDC machine are calculated by using 2-D and 3-D FEM analysis. Therefore, the objective of this chapter is to introduce the developed FEM models and to discuss the very first results of the FEM analysis. In the following sections, the 2-D and 3-D FEM models of the ADR-BLDC machine are described first. Then, the magnetic field distribution results calculated with these models are compared in the base speed range, in order to show the validity of the developed models. In the last part, the dependency of the magnetic field distribution on the axial rotor displacement is analyzed.

4.1 Finite Element Method Analysis

FEM is a numerical approximation method, where partial differential equations are used. In electromagnetic problems, Maxwell's equations and material equations constitute the basis of this analysis. FEM analysis basically involves the following four steps:

- discretization of the solution region into a finite number of elements,
- derivation of the governing equations in an element with the help of a form function,
- assembly of the element equations to generate the system equation,
- solution of the system with respect to the boundary conditions.

The FEM calculations in this study are carried out with the Ansys Maxwell v14 finite element software by using magnetostatic and transient solvers. In Ansys Maxwell, the calculation area is discretized with triangular mesh elements in 2-D calculations and with tetrahedral in 3-D calculations, which permit a flexible discretization of complex geometries and an automated mesh generation. The adaptive meshing is implemented, in order to provide an efficient mesh generation. In order to minimize the discretization error, the mesh is locally refined after each meshing cycle according to the local energy errors as long as the total energy error is larger than the target value defined by the user. In addition to the energy error, the energy difference between the subsequent meshing cycles is evaluated, in order to monitor the convergence. Consequently, the target values of the energy error and the energy difference have a significant effect on the mesh quality. In addition to the adaptive meshing, the maximum size of the mesh elements and the number of the total mesh elements in each part of a simulation model can optionally be defined by the user by changing the mesh definitions.

The magnetic flux path of most of the conventional PM electric machines can be analyzed by two-dimensional field calculations neglecting the axial dependency of electromagnetic parameters. However, if stator and rotor are misaligned, the axial uniformity of the electromagnetic parameters diminishes. Therefore, 2-D models are used for numerical field calculations in the base speed range with axially aligned stator and rotor, whereas 3-D models are used in the mechanical field weakening range.

The model preparation process includes drawing the geometric outline of the model, assignment of the material properties, definition of the current and voltage sources linked to the coils, and specification of the boundary conditions and the design symmetries. 2-D and 3-D FEM analysis models basically go through the same preparation process steps.

4.1.1 2-D Model

The parts and boundaries of the 2-D FEM model are shown in Fig. 4.1. Due to the symmetrical design of the ADR-BLDC machine, only one pole area is modeled with the related symmetry definitions. The left boundary (LB) line and the right boundary (RB) line are specified as odd symmetry boundaries. Additionally, the calculation area, which also includes the surrounding vacuum areas, is restricted by the outer boundary (OB) curve and the inner boundary (IB) curve. On these curves, vector potential $\vec{A}$ is set to zero, which forces the normal component of the magnetic flux density $\vec{B}_n$ along these curves to be zero and restricts the analysis within these boundaries. Therefore, the surrounding vacuum areas are chosen to be large enough. A band object covering the rotating parts is needed to define the rotational motion of the rotor in the transient analysis. The integral parameters such as flux linkage, induced voltage and inductance values are calculated from the local parameters by using a multiplication factor called model depth, which stands for the axial length of the electric machine l_z. On the right side of Fig. 4.1, the result of the adaptive meshing is shown. As can be seen, the mesh density in the air-gap is higher than the other parts due to the important role of the air-gap field for magnetic energy and torque calculations. Moreover, the stator yoke and the stator teeth are modeled separately, and the maximum length of the mesh elements in the teeth is set to be shorter, in order to increase the accuracy of the numerical calculations.

In a 2-D analysis, the magnetic field intensity $\vec{H}$ and the magnetic flux density $\vec{B}$ have only x and y components. Therefore, the properties of the materials are specified only in

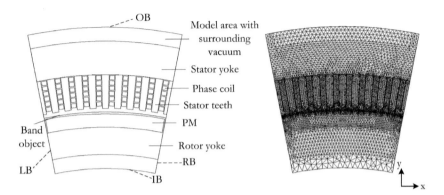

Figure 4.1: 2-D FEM analysis model

the x-y plane. The characteristics of the electrical steel 10JNEX900 are defined by using the magnetization curve shown in Fig. 3.5 according to the lamination model introduced in [77]. This lamination model considers the influence of the laminations between the steel sheets on the equivalent magnetic relative permeability of the electrical steel in the x-y plane $\mu_{r,xy}^*$ as given by

$$\mu_{r,xy}^* = (1 - k_{lam}) + k_{lam}\mu_{r,fe} , \qquad (4.1)$$

where $\mu_{r,fe}$ is the relative permeability of the electrical steel, and k_{lam} is the stacking factor. The stacking factor is defined as the ratio of the total axial length of the steel l_{fe} to the total axial length of the stator stack as given by

$$k_{lam} = \frac{l_{fe}}{l_{fe} + l_0} , \qquad (4.2)$$

where l_0 is the total axial length of the laminations. As given in [72], the stacking factor of the electrical steel 10JNEX900 is $k_{lam} = 0.9$. After the equivalent permeability is determined, the original axial length is used as the model depth l_z in the 2-D model. The rotor steel C45 is modeled with its B-H curve given in Fig. 3.5. The demagnetization curve of the PMs is modeled with the constant value of the residual flux density B_r depending on rotor temperatures according to the data given in [78]. Additionally, the physical properties of the materials are modeled according to the data given in table 3.1.

4.1.2 3-D Model

The 3-D FEM model is shown in Fig. 4.2, where the model volume, the band object volume, the left and right symmetry boundary planes, and the electric machine parts are marked. Similar to the 2-D model, just one pole area is modeled. Assuming that the electric machine has two axially symmetrical sides, only one axial half of the ADR-BLDC machine is modeled, in order to reduce the calculation time. The right boundary (RB) and left boundary (LB) planes are defined as odd boundaries. Except for the symmetry boundaries, the other boundary definitions are not required in 3-D models, since the faces of the calculation volume are automatically treated as Neumann boundaries, where the normal component of the magnetic field intensity $\vec{H}_n$ is set to zero. Therefore, the outer semicircular, front and back faces of the model volume are automatically treated as Neumann boundaries. Additionally, the end winding turns of the ADR-BLDC machine are implemented in this model. This is an important feature of the 3-D model, since the end winding conductors have to be modeled due to the fact that they participate in torque generation. In the bottom image of the same figure, the result of the adaptive meshing with an aligned stator and rotor is shown. The mesh options have to be optimized in 3-D models because of the trade-off between accuracy and the computational effort.

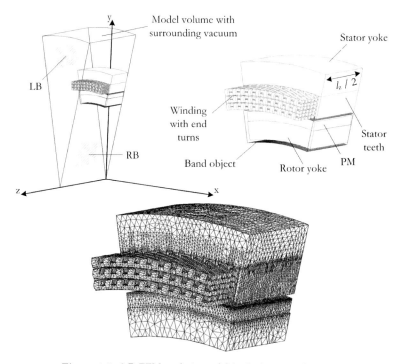

Figure 4.2: 3-D FEM analysis model in the base speed range

The anisotropic characteristics of the materials have to be defined in 3-D models. Since the rotor material steel C45 and the PM material BM 38SH are isotropic, the same characteristics used in the 2-D model are used. On the other hand, the stator material has anisotropic characteristics due to the laminations. The characteristics of the electrical steel in the stator's x-y plane are defined similarly to the 2-D model. Additionally, the equivalent magnetic permeability of this material in the z direction is needed. This can be determined by using the magnetic reluctances [72].

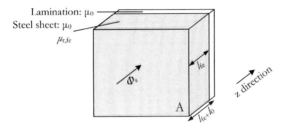

Figure 4.3: Steel sheet and lamination part of electrical steel

According to the information given in Fig. 4.3, the magnetic reluctances of the steel $R_{m,fe}$ and the lamination $R_{m,0}$ can be calculated by

$$R_{m,fe} = \frac{l_{fe}}{\mu_0 \mu_{r,fe} A} \tag{4.3}$$

and

$$R_{m,0} = \frac{l_0}{\mu_0 A} \, . \tag{4.4}$$

By equating the sum of these magnetic reluctances to the equivalent reluctance of the electrical steel as in

$$\frac{l_{fe} + l_0}{\mu_0 \mu_{r,z}^* A} = \frac{l_{fe}}{\mu_0 \mu_{r,fe} A} + \frac{l_0}{\mu_0 A} \, , \tag{4.5}$$

the equivalent relative permeability of the electrical steel in the z direction $\mu_{r,z}^*$ is found as

$$\mu_{r,z}^* = \frac{l_{fe} + l_0}{l_{fe}/\mu_{r,fe} + l_0} = \frac{\mu_{r,fe}}{\mu_{r,fe} - k_{lam}(\mu_{r,fe} - 1)} \, . \tag{4.6}$$

Fig. 4.4 shows the equivalent relative permeabilities of the electrical steel in the stator both in the x-y plane (Eqn. 4.1) and in the z direction (Eqn. 4.6). As can be seen, $\mu_{r,z}^*$ is almost constant and much lower than $\mu_{r,xy}^*$, unless the material is highly saturated.

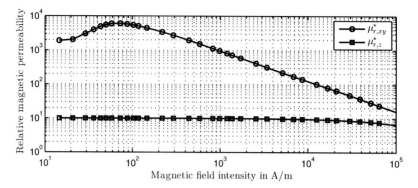

Figure 4.4: Equivalent relative magnetic permeability characteristics in the x-y plane and in the z direction

4.2 Comparison of the 2-D and 3-D FEM Models

The results of the 2-D and 3-D FEM analyses are compared, in order to verify the validity of the 2-D model in the base speed range and to test the adequacy of the discretization of the 3-D model. It is expected that the results of the 2-D and 3-D FEM models are similar in the base speed range, except for the end effects such as the fringing fields and the influence of the end winding. Therefore, the magnetic field distributions and the torque calculated by these simulation models are compared with an axially aligned stator and rotor.

Two load cases, where all phase currents are equal to zero and the eight phases are supplied with 25 A, are simulated by both models. The average torque calculated from the 2-D analysis with 25 A is equal to 117.8 Nm, which matches well with 118.4 Nm from the 3-D model.

Fig. 4.5 shows the distributions of the magnitude of the magnetic flux density B_{mag} simulated by both models. As can be seen, the rotor magnetic flux density calculated by 3-D FEM analysis is homogeneous in the axial direction, and the 2-D and 3-D models deliver similar results. However, the stator magnetic flux density distribution calculated by 3-D FEM analysis is inhomogeneous in the axial direction because of the end effects. Due to the symmetry axis defined at the back plane of the 3-D model, the magnetic flux density distributions on this plane, which are shown in the 3-D model back views, match with the 2-D analysis results. Another difference between these results is the shape of the magnetic flux density distribution in the stator yoke. In the 2-D results, the low flux density area has a triangular shape compared to the circular shape in the 3-D results. This difference arises from the lower number of mesh elements in this part of the 3-D model and the different interpolation methods used by the 2-D and 3-D solvers. Its influence on the results of the FEM analysis is negligible, since it occurs at low magnetic field density region, and this part of the stator does not have a remarkable effect on the end results.

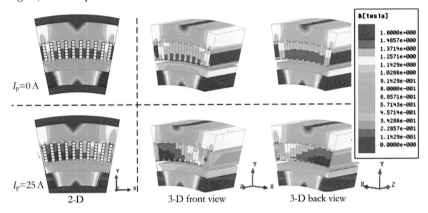

Figure 4.5: Distribution of the magnitude of magnetic flux density calculated in the base speed range without phase currents (top) and eight stator phases being supplied with 25 A (bottom) by 2-D and 3-D FEM magnetostatic analyses; Front view shows the axial end of the stator core; Back view shows the axial middle of the stator core

In addition to the visual comparisons, the magnetostatic FEM analysis results of the 2-D and 3-D models are compared quantitatively by evaluating the magnetic flux density values in the stator yoke, stator teeth, PM and rotor yoke. Fig. 4.6 shows the evaluation points in the 2-D model and the evaluation lines in the 3-D model.

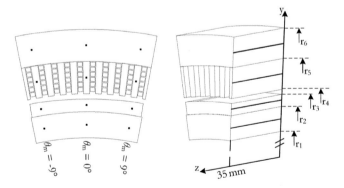

Figure 4.6: Evaluation points in the 2-D model and evaluation lines in the 3-D model

The diagrams in Fig. 4.7 show the numerical calculation results without phase currents. In these diagrams, the magnetic flux density values from the 2-D model are kept constant along the z direction to be able to make a comparison. Both models deliver comparable results in all machine parts, except for the above mentioned small difference in the stator yoke. Moreover, the end effects are seen in the 3-D analysis results in the stator teeth and PM, while they can be ignored in the stator and rotor yokes. It is important to note that the magnetic flux density distribution is symmetrical with respect to the pole middle, since flux distribution is caused only by the PM flux at no-load. For this reason, B_{mag} has the same values at $\theta_{\mathrm{m}} = -9°$ and $\theta_{\mathrm{m}} = 9°$ in all the machine parts.

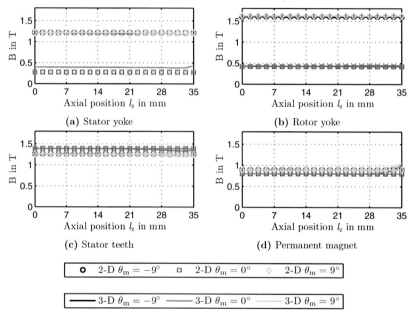

Figure 4.7: Magnitude of the magnetic flux density without phase currents

Fig. 4.8 shows the numerical calculation results with the stator phases being supplied with 25 A. It is also seen in these diagrams that the 2-D and 3-D analysis results match well with each other. The symmetrical distribution of the magnetic flux density in the stator yoke and teeth diminishes with the phase currents. The end effects in the stator differ from the case without phase currents mainly due to the current distribution in the end winding conductors.

It is evident from these results that both models deliver comparable results, except for the end effects. Therefore, the ADR-BLDC machine can be numerically analyzed with the

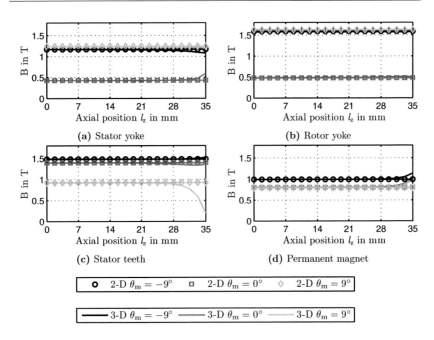

Figure 4.8: Magnitude of the magnetic flux density, the eight phases being supplied with 25 A

2-D FEM model in the base speed range. Nevertheless, the end effects due to the finite axial length of the electric machine and the end winding related parameters can only be computed by the 3-D FEM analysis model. The analysis in this part also shows that the discretization of the 3-D model is sufficient, and it is feasible to model only one rotor part side of the ADR-BLDC with the appropriate symmetry definition.

4.3 Influence of the Axial Rotor Displacement on the Magnetic Field Distribution

In this section, the influence of the axial rotor displacement on the magnetic field distribution is examined by 3-D FEM analysis. The aim of these analyses is to highlight possible changes in the ADR-BLDC machine characteristics in the mechanical field weakening range.

The terminology used and the geometrical definitions of the ADR-BLDC machine are shown in Fig. 4.9 on the same y-z plane defined in Fig. 4.6. The axial stator/rotor overlap length $l_{\mathrm{s-r}}$ represents the axial length of the electric machine, where the stator stands axially over the rotor whereas the rotor overhang is the rotor part that stands axially beyond the stator edge.

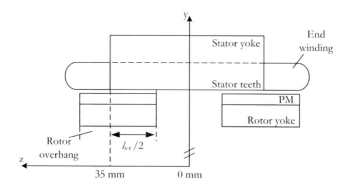

Figure 4.9: Terminology and geometrical definitions

Fig. 4.10 shows the magnetic flux density distributions in the y-z plane at $l_{\mathrm{s-r}} = 40\,\mathrm{mm}$ and $l_{\mathrm{s-r}} = 24\,\mathrm{mm}$ without phase currents. From these results, the following can be concluded:

- The magnetic flux density is inhomogeneous in the axial direction not only due to the end effects but also due to the stator/rotor misalignment. Therefore, the magnetic flux in the machine parts is composed of z-components in addition to the components in the x-y plane.

- As the rotor sections are axially displaced, the PM flux in the axial center of the stator diminishes, and it concentrates in the outer stator regions. Therefore, the phase flux linkage reduces, as the axial stator/rotor overlap length decreases. This is desired, since it allows a wider extended speed range with a constant DC-link voltage without requiring a negative d-axis current.

• The magnetic flux is not only present in the stator and rotor cores but also in the other machine parts, such as the end winding and the construction parts, mainly because of the magnetic flux caused by the rotor overhang.

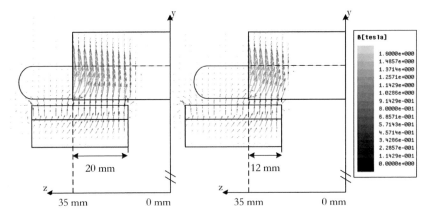

Figure 4.10: Magnetic flux density vector in the y-z plane at $l_{s-r} = 40\,\text{mm}$ and $l_{s-r} = 24\,\text{mm}$ without phase currents

As a result of the discussions above, the machine parameters of the ADR-BLDC machine such as the flux linkage and back EMF of the phases and the torque constant change, as the rotor is axially displaced. Moreover, the inductance values are expected to change with the axial rotor position due to the changes in the local permeability values in the stator core. Due to the changes in the machine parameters, the operational characteristics of ADR-BLDC machines are also expected to change. The influence of the axial rotor displacement on the electrical machine parameters and the operating characteristics of the ADR-BLDC machine are analyzed in the following two chapters in detail.

5 Machine Parameters

Electric machine parameters, such as the back EMF, torque constant, flux linkage and inductances, describe the characteristic of electric machines and are needed for the implementation of control algorithms. These parameters are calculated with analytic and numeric calculations in the design process, and they typically depend on the phase currents, rotational speed and rotor angular position. Conventional calculation methods are applicable for ADR-BLDC machines in the base speed range. However, the parameters of the electric machine differ from those in the base speed range during the mechanical field weakening operation. Moreover, the dependency of the machine parameters on the axial rotor position specifies the operational limits of the mechanical field weakening and displays the changes required in the control algorithm during the mechanical field weakening operation. Therefore, the influence of the axial rotor displacement on the machine parameters of the ADR-BLDC machine is examined in this chapter. The back EMF waveforms, the cogging torque, the torque constant and the inductance values of the electric machine are determined depending on the axial rotor displacement. Additionally, the measured back EMF waveforms and the no-load inductances are used to validate the calculation results.

5.1 Back EMF

First of all, the phase flux linkage Ψ_{PM} waveforms of the ADR-BLDC machine at varying axial rotor positions are calculated by 3-D FEM analysis. These waveforms are shown in Fig. 5.1, where a decrease in the amplitude of the flux linkage with a decreasing axial stator/rotor overlap length l_{s-r} is seen. Nevertheless, the dependency of Ψ_{PM} on the axial rotor position is not seen directly from these results. Therefore, the effect of the axial rotor displacement on the amplitude of the flux linkage is separately illustrated in Fig. 5.2. The amplitudes of the flux linkage without stator/rotor misalignment effects are additionally illustrated in this figure. At $l_{s-r} = 70\,\mathrm{mm}$, the rotor is aligned with the stator, and these values are equal. As l_{s-r} decreases, these values differ from each other. It can be concluded that the phase flux linkage due to the PM flux is affected by the axial rotor displacement, but it cannot be assumed to decrease linearly with l_{s-r}.

In order to explore the reasons for the increase in the flux linkage values compared to the results of the 2-D field distribution case, the magnetic flux density distributions in the mechanical field weakening range are analyzed in detail. First of all, the air-gap magnetic

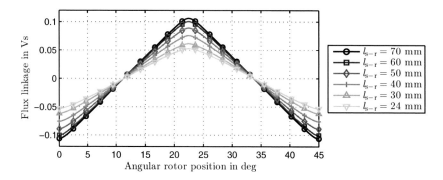

Figure 5.1: Flux linkage waveforms of a phase at varying axial stator/rotor overlap lengths l_{s-r} calculated by 3-D FEM analysis

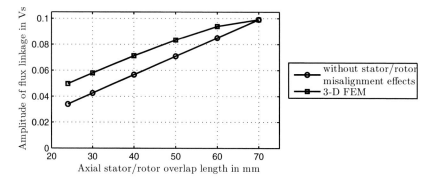

Figure 5.2: Amplitude of the flux linkage waveform calculated by 3-D FEM analysis and amplitude of flux linkage waveform determined without the stator/rotor misalignment effects

flux density distributions are analyzed. The distributions of the radial flux density B_{rad} along the axial axes in the middle of the air-gap are illustrated in Fig. 5.3 at $l_{s-r} = 70$ mm (axially aligned stator and rotor) and $l_{s-r} = 40$ mm. In both of these plots, B_{rad} has an odd symmetry at any axial position over an electrical period, and the influence of the stator teeth is seen. When the stator and the rotor are axially aligned, the B_{rad} distribution is homogeneous in the axial direction except for the end effects, which cause a decrease in the B_{rad} values for $l_z > 27$ mm. In the mechanical field weakening range, B_{rad} is zero or negligibly small in the middle of the stator between 0 mm $< l_z < 7.5$ mm. The B_{rad} values are non-zero at $l_z > 7.5$ mm because of the inner fringing effects, even though the rotor stays axially between $15\,\mathrm{mm} < l_z < 50\,\mathrm{mm}$. They arise due to the higher relative permeability of the inner part of the electrical steel in the stator, which is exposed to a lower PM flux through the air-gap than the outer stator regions. Moreover, the decrease in B_{rad} due to the end effects is lower, since the rotor overhang flux also contributes to the air-gap flux. Apart from these differences in the rotor and stator end plate regions, the flux density distribution in the air-gap is comparable in both cases. Consequently, the air-gap flux per axial length ($\Phi_g/l_{\mathrm{s-r}}$) increases as $l_{\mathrm{s-r}}$ decreases due to the inner fringing effects and the overhang flux contributing to the air-gap flux. This is one of the reasons for the difference between the linear approximation and the 3-D results in Fig. 5.2.

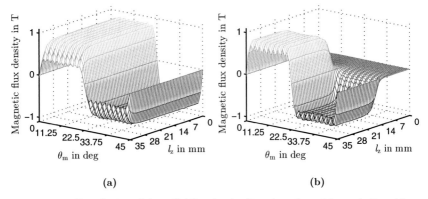

(a)   (b)

Figure 5.3: Distribution of the radial flux density B_{rad} along the axial axes in the middle of the air-gap a) $l_{\mathrm{s-r}} = 70\,\mathrm{mm}$ and b) $l_{\mathrm{s-r}} = 40\,\mathrm{mm}$

In addition to the dependency of the air-gap magnetic flux on the axial rotor displacement, the rotor overhang causes an additional flux in the end regions of the stator as well as flux components that are not confined in the stator and rotor cores. These flux components affect the phase flux linkages. To visualize them, the magnetic flux density distributions in the y-z plane at $l_{\mathrm{s-r}} = 70\,\mathrm{mm}$ and $l_{\mathrm{s-r}} = 40\,\mathrm{mm}$ are shown in Fig. 5.4, where the flux paths are marked according to the flux components in the stator. It is important to note that the end effects due to finite axial length of stator and rotor are excluded for simplicity reasons. The first flux path represents the main flux of the ADR-BLDC machine. The 2nd, 3rd and 4th flux components result from the stator/rotor misalignment in the mechanical field weakening range. The inner fringing effects are shown by the second flux path. The flux lines of the third flux path penetrate the end winding and enter the stator at its axial end. This axial flux is directed radially in the stator core because of the higher permeability of the laminated steel material in this direction. Finally, the fourth flux path shows the overhang fringing flux.

As can be seen from the right plot in Fig. 5.4, the phases link the additional flux components not only in the stator core but also in the end winding regions. Consequently, the additional

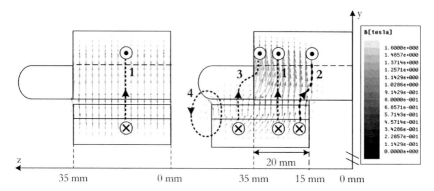

Figure 5.4: Flux density distribution in the x-y plane at a) $l_{s-r} = 70\,\text{mm}$ and b) $l_{s-r} = 40\,\text{mm}$

flux components causing the difference in Fig. 5.2 can be divided as follows:

- additional flux in the stator core (2nd and 3rd flux components),
- flux which is linked in the end winding regions (3rd and 4th flux components).

The increase in the amplitude of the flux linkage depends on both the additional flux components as well as the winding distribution in the stator slots and in the end winding region. Accordingly, the configuration of the end winding turns has a considerable effect on the flux linkage at low l_{s-r}. The resulting flux linkage of a phase can be split into three components (air-gap, stator additional and end winding flux linkage components), as given in Eqn. 5.1. The resulting flux linkage is a function of the rotor angular position θ_m and the axial stator/rotor overlap length l_{s-r} as well as of the machine design.

$$\Psi(\theta_m, l_{s-r}) = \Psi_{\text{air-gap}}(\theta_m, l_{s-r}) + \Psi_{\text{stator axial}}(\theta_m, l_{s-r}) + \Psi_{\text{end winding}}(\theta_m, l_{s-r}) \qquad (5.1)$$

In order to separate the stator additional and end winding flux linkage components, the part of the flux linkage due to the flux components only in the stator core is calculated. These data are obtained by using the magnetic field distribution results of 3-D magnetostatic FEM analysis. The radial and circumferential winding distributions are taken into account, since the stator axial flux varies also in radial direction. These results are marked in Fig. 5.5 with diamonds. Accordingly, the dark gray shaded area in this figure represents the part of the flux linkage due to the stator additional flux and the light gray shaded area the part of the flux linkage due to the flux linked in the end winding regions. As can be seen in this figure, the additional flux components in the stator core and end winding regions must be considered in the back EMF calculations, since they constitute almost one third of the total flux linkage at $l_{s-r} = 24\,\text{mm}$.

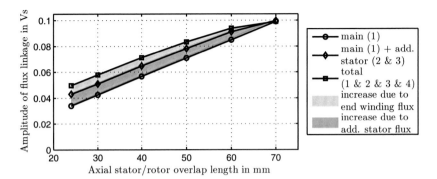

Figure 5.5: Effects of additional flux components in the stator core and in the end winding regions on the amplitude of the flux linkage

The amplitude of the back EMF waveforms has the same dependency on the axial rotor position as the flux linkage. The back EMF waveforms at $n = 1000\,\mathrm{rpm}$ are calculated by transient 3-D FEM analysis, and the results are shown in Fig. 5.6. In this analysis, the influence of the induced eddy currents on the magnetic field distribution is neglected. As can be seen, the induced voltage is trapezoidal, if stator and rotor are axially aligned. As $l_{\mathrm{s-r}}$ decreases, the back EMF waveforms are getting less trapezoidal. Indeed, their form becomes a combination of trapezoidal and triangular waves at low $l_{\mathrm{s-r}}$. This is caused by the end winding effects. The end winding configuration and the magnetic field distribution in the end winding regions determine this dependency. The triangular back EMF part emerges, since the ADR-BLDC machine has a triangular end winding form. In Fig. 5.7, the influence of the additional flux components on the back EMF characteristic is visualized for the minimum $l_{\mathrm{s-r}} = 24\,\mathrm{mm}$. The thin line on this graph shows the back EMF waveform without the influence of the additional flux components, the line in the middle shows the back EMF waveform, considering the additional flux components in the stator core. Finally, the thick line shows the resulting back EMF waveform. As can be seen in this figure, the first two curves have the same waveform, but the end winding effects change this trapezoidal form.

The amplitude of the back EMF waveform can be varied by changing the axial stator/rotor overlap length. Nevertheless, the effectiveness of this method is limited by the additional flux components in the stator core and in the end winding regions. If the stator/rotor overlap length is decreased to 35% of its original value, the amplitude of the back EMF waveform decreases by 50%. Moreover, the characteristic of the back EMF waveform changes due to the influence of the end winding, if mechanical field weakening is applied.

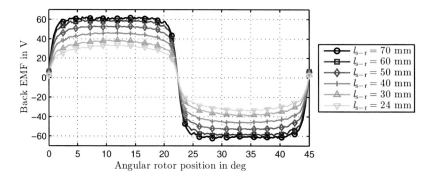

Figure 5.6: Back EMF waveforms at 1000 rpm calculated by 3-D FEM analysis

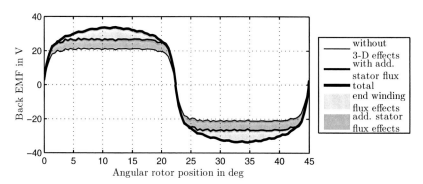

Figure 5.7: Components of the back EMF waveform at 1000 rpm and $l_{s-r} = 24$ mm calculated by 3-D FEM analysis

5.2 Torque Characteristics

In an electric machine, the forces act on

- current carrying conductors in a magnetic field,
- interfaces between flux conducting materials with different magnetic permeabilities [67].

The torque is the consequence of the circumferential force and leads to a mechanical rotation. Therefore, the forces in the tangential direction are required to cause an electromechanical energy conservation. In addition to the tangential forces, there are forces in axial and normal direction. These may cause mechanical oscillations and mechanical stress on the construction parts, thus they must be considered also in the design process.

As discussed in section 2.2.2, the average torque produced by a PM brushless machine is composed of the synchronous and reluctance torque components. In addition, there are undesired torque ripple components with an average value equal to zero. The first one is the cogging torque, which arises due to the interaction between rotor PM flux and variable air-gap reluctance as a result of slotting. This torque component is independent of the stator currents, hence it is measured and calculated with the phases being open circuited. If the stator currents are different from zero, the torque ripple is called pulsating torque, which is caused by the interaction between rotor PM flux and variable air-gap reluctance as well as MMFs due to the stator currents.

In this section, first the torque characteristics of the ADR-BLDC machine in the base speed range and then the influence of the axial rotor displacement on the torque characteristics in the mechanical field weakening range are examined.

Base Speed Range

The torque produced by the ADR-BLDC machine in the base speed range is calculated, assuming the phase currents have ideal characteristics with an 80% conduction duration by using the 2-D FEM analysis model. The ideal current waveform of a phase depending on the rotor position is illustrated in Fig. 3.7. The amplitude of the torque oscillation and the average torque are shown in Fig. 5.8 as a function of the amplitude of the phase current.

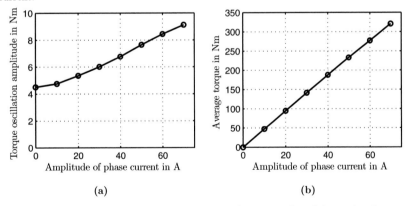

| (a) | (b) |

Figure 5.8: Amplitude of the torque oscillation and average value of the produced torque as a function of the amplitude of the stator phase current

The amplitude of the cogging torque is less than 3% of the rated torque (150 Nm). Although measures, such as semi-closed slots and skewed stator or rotor designs, are not taken, the cogging torque of the ADR-BLDC machine is in an acceptable range. This is because of the fact that SMPM rotor BLDC machines have lower cogging torque values due to their relatively large effective air-gap compared to IPM rotor machines. Fig. 5.9 shows the cogging torque of the ADR-BLDC machine for an electrical period. Since the ratio of the number of slots to the number of poles S/n_{p} of the ADR-BLDC machine is an integer, there are S=160 cogging cycles per mechanical revolution and $S/n_{\mathrm{p}} = 20$ cogging cycles per electrical period.

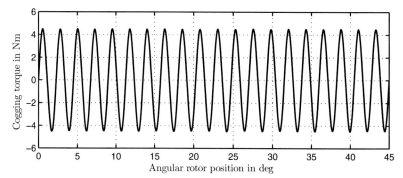

Figure 5.9: Waveform of the cogging torque over an electrical period

As can be seen in Fig. 5.8a, the amplitude of the pulsating torque increases with the amplitude of the phase current. However, it is not critical, since the amplitudes of the pulsating torque are less than 3% of the corresponding average torque values.

The average torque diagram in Fig. 5.8b shows that the produced torque varies almost linearly with the phase current amplitude. Actually, the torque constant $k\Phi$ decreases as the net produced torque increases due to the saturation of the stator material. However, it differs with phase currents less than 2.5%. Therefore, the torque constant $k\Phi$ is taken as equal to 0.55 Nm/A under the above defined conditions.

If the stator phases are supplied with block shape phase currents with 80% conduction duration, the input DC current I_{dc} in Eqn. 3.20 is equal to eight times the amplitude of the phase current $\hat{i}_{\mathrm{p}}$, since eight out of ten phases are conducting at any time. Consequently, the average torque produced by the ADR-BLDC machine in the base speed range can be calculated by

$$T = k\Phi I_{\mathrm{dc}} = 0.55 \times 8 \times \hat{i}_{\mathrm{p}}\,. \tag{5.2}$$

Mechanical Field Weakening Range

First of all, the cogging torque of the ADR-BLDC machine is analyzed by using 3-D FEM analysis. The amplitude of the cogging torque is plotted against the axial stator/rotor overlap length l_{s-r} in Fig. 5.10. At $l_{s-r} = 70\,\text{mm}$, stator and rotor are axially aligned, and the cogging torque calculated with 3-D FEM analysis is comparable with the 2-D result given in Fig. 5.8a. As l_{s-r} decreases, the amplitude of the cogging torque decreases, since the effective air-gap between stator and rotor increases. This decrease is not linear with l_{s-r} due to the additional flux components.

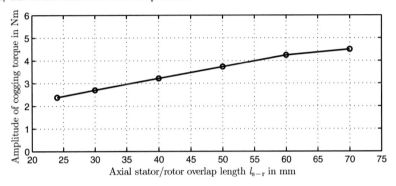

Figure 5.10: Amplitude of the cogging torque as a function of l_{s-r} simulated by 3-D FEM analysis

The pulsating torque at a given axial rotor position is expected to increase with the phase currents similarly to the aligned stator and rotor case. Some differences between these two cases are expected due to the saturation of the stator teeth caused by the axial flux components. This dependency of the pulsating torque on the phase currents is not examined further in this study due to its relatively small amplitude, which does not have a crucial influence on the machine performance.

Due to the additional flux components in the stator core and in the end winding region, the torque of the ADR-BLDC machine must be calculated by using 3-D FEM analysis. However, instead of carrying out additional 3-D analysis calculations, the known phase flux linkage characteristics from Fig. 5.1 can be used to calculate an approximated average torque value. This calculation can be carried out by using

$$T = \sum_{k=1}^{10} \left(\frac{1}{2\pi} \int_0^{2\pi} i_k(\theta_r) \, \frac{\mathrm{d}\Psi_{\mathrm{PM},k}(\theta_r, l_{s-r})}{\mathrm{d}\theta_r} \, \mathrm{d}\theta_r \right), \tag{5.3}$$

which is derived by writing Eqn. 3.19 in terms of the flux linkage according to Eqn. 3.10 and taking the average of the instantaneous torque of a phase over an electrical period.

Eqn. 3.10 is used to determine the resultant average torque values, provided that the phase currents have ideal characteristics. Additionally, some exemplary load conditions are simulated directly with 3-D FEM analysis, in order to validate these results. Fig. 5.11 shows the resulting average torque values both from Eqn. 5.3 and from 3-D FEM analysis with phase currents. As can be seen, both calculation methods deliver very close results. This indicates that the flux linkage characteristics calculated without phase currents are sufficient to determine the average torque of the ADR-BLDC machine. Moreover, since the average torque at a given axial rotor position changes almost linearly with the amplitude of the phase current, the torque coefficient of the ADR-BLDC machine can be assumed to be constant at a given axial rotor position. Accordingly, the torque constant of the ADR-BLDC machine is shown in Fig. 5.12 as a function of the axial rotor position. It is important to note that these results are only valid for phase currents having ideal characteristics. In case of other phase current characteristics, the saturation effects must be considered again.

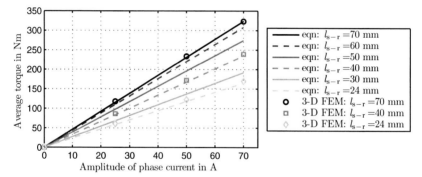

Figure 5.11: Average torque as a function of the amplitude of the stator phase current calculated by using Eqn. 5.3 at varying l_{s-r} values and compared with 3-D FEM analysis results

After determining the rotor position dependency of the machine torque constant, the torque produced by the rotor overhang is examined. The rotor of the BLDC machine is divided into the inner part and the rotor overhang in the 3-D FEM model, as shown in Fig. 5.13. The torque produced by each of these parts is evaluated separately, and the results are given in table 5.1. As can be seen, the torque produced due to the inner rotor part changes linearly with l_{s-r}, but the torque produced due to the rotor overhang increases slightly as the overhang length increases. Moreover, the rotor overhang torque is lower than the inner rotor torque. This is caused due to the following reasons:

- The flux density in the end winding region is lower compared to the stator core.
- The local flux density vectors in the end winding region do not only have tangential but also axial components, which do not participate in the torque production.

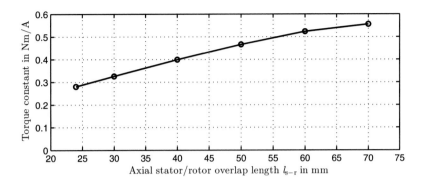

Figure 5.12: Torque coefficient of the ADR-BLDC machine as a function of l_{s-r}

- The end winding conductors are not perpendicular to the rotation axis, so the force on the end winding conductors also has an axial component, which does not participate in the torque production.

However, one third of the total torque is produced by the rotor overhang, if l_{s-r} is minimum. This indicates that the rotor overhang torque must be considered because of its high share in the resultant torque at low l_{s-r}.

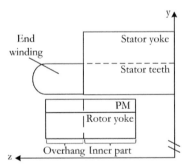

Figure 5.13: Rotor parts in the 3-D FEM model illustrated in the y-z plane

The active axial length of the ADR-BLDC machine can be defined by using the results for the machine torque constant. The torque constant values are more suitable for defining the active axial length than the amplitude of phase flux linkages, since the machine torque constant is related to the average value of the phase flux linkage over a phase conduction duration (Eqn. 5.3). The ideal machine length l_i at the base position is taken equal to the axial stack length l_z due to the fact that the axial lengths of the stator core and the

$\hat{i}_{\mathrm{p}}$	$l_{\mathrm{s-r}} = 40\,\mathrm{mm}$		$l_{\mathrm{s-r}} = 24\,\mathrm{mm}$	
	T_{inner}	T_{overhang}	T_{inner}	T_{overhang}
25 A	67.08 Nm	19.35 Nm	40.21 Nm	21.07 Nm
50 A	133.67 Nm	38.59 Nm	80.25 Nm	42.20 Nm
70 A	185.50 Nm	53.94 Nm	111.25 Nm	58.77 Nm

Table 5.1: Produced torque of the inner rotor part T_{inner} and the rotor overhang T_{overhang}

rotor are the same and the effective air-gap of the electric machine is relatively large. Fig. 5.14 shows the active machine length as a function of $l_{\mathrm{s-r}}$. Additionally, the active machine lengths at lower $l_{\mathrm{s-r}}$ than 24 mm are illustrated with a dashed line in this figure. As can be seen, the rate of decrease of the active axial length is lower at $l_{\mathrm{s-r}}$ values close to $l_{\mathrm{z}} = 70\,\mathrm{mm}$ than low $l_{\mathrm{s-r}}$ values. For $l_{\mathrm{s-r}} \leq 50\,\mathrm{mm}$, the active axial length decreases almost with a constant slope. This shows that the inner fringing effects and the rotor overhang flux limit the decrease in the active axial length, but these effects stay almost constant at relatively low $l_{\mathrm{s-r}}$ values. However, this behavior changes slightly, if $l_{\mathrm{s-r}}$ is reduced almost to zero. As the flux density in the outer regions of the stator is low, the inner fringing effects and the rotor overhang flux in the stator core increase slightly. These results also show that the active axial length of the ADR-BLDC machine can be reduced to 14 mm by reducing $l_{\mathrm{s-r}}$ to zero.

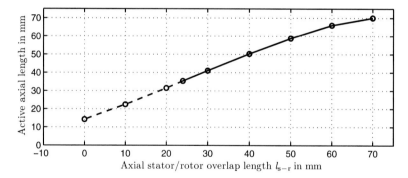

Figure 5.14: Active machine length as a function of $l_{\mathrm{s-r}}$

5.3 Phase Inductance

Phase inductances are amongst the important machine parameters because of their effect on the operational behavior of electric drives. The inductance values of the ADR-BLDC machine require special consideration because of the high number of stator phases and the magnetic couplings between them.

In this section, first the inductances of the ADR-BLDC machine are examined with axially aligned stator and rotor, and their dependency on the angular rotor position and the phase currents is briefly discussed. Then, the influence of the axial rotor displacement on the phase inductances is analyzed. Finally, a numeric calculation method that is used to determine the inductance values in the mechanical field weakening range by using 2-D FEM and simplified 3-D FEM analyses is proposed.

5.3.1 Base Speed Range

Phase inductances can analytically be estimated by assuming pre-defined flux paths and ignoring the saturation of the ferromagnetic materials. Based on these assumptions, the approximated magnetic field distribution as a result of the phase current in a slot opening is shown in Fig. 5.15a. In this figure, only the leakage magnetic field of a phase coil is considered. The magnetic field in the slot is called the slot leakage field and the magnetic field in the air-gap is called the air-gap leakage field. The resulting magnetic field must be approximated by additionally considering the PM field.

Since the magnetic field distribution in a slot can be approximated well, the inductance of the phase due to the slot leakage flux can be estimated by calculating the magnetic energy stored in a slot. In this calculation, the magnetic energy stored in the stator core is neglected due to $\mu_{\text{fe}} >> \mu_0$. Moreover, the current distribution in the slot is assumed to be homogeneous.

Under these assumptions, the slot leakage flux density at a given height x can be calculated on the defined integration path in Fig. 5.15a by

$$B(x) = \mu_0 \, H(x) = \mu_0 \, \frac{z(x)\, i}{b} \, . \tag{5.4}$$

In this equation, $z(x)$ is the number of the conductors as a function of x. Since only the coil sides of a phase are placed into a slot, $z(x)$ can be defined as

$$z(x) = n_{\text{c}} \, \frac{x}{h} \, . \tag{5.5}$$

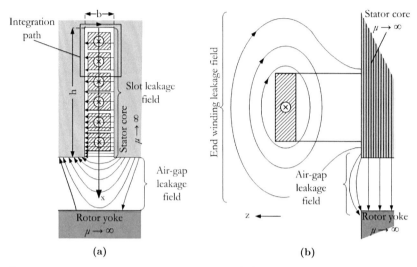

Figure 5.15: Leakage magnetic field of a phase coil [67], where the PM is not considered, since $\mu_{\mathrm{PM}} \approx \mu_0$

$z(x)$ increases linearly with x since the slot opening width of the ADR-BLDC machine is constant. Accordingly, the stored magnetic energy in the slot can be written as

$$
\begin{aligned}
W_\Phi &= \frac{1}{2\mu_0} \int B(x)^2 \, \mathrm{d}V = \frac{1}{2} \int \frac{\mu_0 \, z(x)^2 \, i^2}{b^2} \, \mathrm{d}V \\
&= \frac{1}{2} \int \frac{\mu_0 \, n_{\mathrm{c}}^2 \, x^2 \, i^2}{b^2 \, h^2} \, \mathrm{d}V .
\end{aligned}
\tag{5.6}
$$

By substituting the volume element $\mathrm{d}V$ by

$$
\mathrm{d}V = l_z \, b \, \mathrm{d}x,
\tag{5.7}
$$

the stored magnetic energy in a slot is written by

$$
W_\Phi = \frac{\mu_0 \, n_{\mathrm{c}}^2 \, i^2 \, l_z}{2 \, b \, h^2} \int\limits_0^h x^2 \, \mathrm{d}x = \frac{\mu_0 \, n_{\mathrm{c}}^2 \, i^2 \, l_z \, h}{6 \, b} .
\tag{5.8}
$$

By multiplying the stored magnetic energy by the number of slots per phase, which is equal to $2 \, n_{\mathrm{p}} = 16$, and equating this to the stored magnetic energy in an inductance given as

$$W_{\mathrm{L}} = \frac{1}{2} L\, i^2,$$ (5.9)

the slot leakage inductance results in

$$L_{\mathrm{s}} = \frac{2\, n_{\mathrm{p}}\, \mu_0\, n_{\mathrm{c}}^2\, l_z\, h}{3\, b} = 89.7\,\mu\mathrm{H}.$$ (5.10)

The air-gap leakage inductance L_{g} can be determined similarly, if the magnetic field distribution in the air-gap is known. However, it is not possible to approximate the flux paths in the air-gap as in a stator slot.

In order to find out the range of L_{g}, the inductance of a stator phase is calculated by using the 2-D FEM model, where the relative permeabilities of the stator and rotor cores are fixed to a very high value equal to 10000 and the PM material is defined as vacuum, so that the saturation effects and the PM flux are not included in these simulations. The simulated total phase inductance is independent of the axial rotor position, and it is equal to 197 μH. Moreover, from the stored magnetic energy in the slot, the slot leakage inductance is calculated $L_{\mathrm{s}} = 82\,\mu\mathrm{H}$, which agrees very well with the resulting in analytical calculation result given in Eqn. 5.10. From these results, L_{g} is found to be 115 μH. Consequently, L_{s} and L_{g} values are in the same range.

The last component of the stator phase inductance is the end winding leakage inductance L_{e}. According to [79], the end winding inductance of the ADR-BLDC machine can be estimated by

$$L_{\mathrm{e}} = \frac{n_{\mathrm{p}}\mu_0\tau_{\mathrm{cp}}n_{\mathrm{c}}^{\,2}}{2} ln\left(\frac{\tau_{\mathrm{cp}}\sqrt{\pi}}{\sqrt{2A_{\mathrm{s}}}}\right) = 16.4\,\mu\mathrm{H},$$ (5.11)

where the approximated diameter of the end turns τ_{cp} is taken as 40 mm and the slot cross-sectional area A_{s} equals to 27.1 mm. This inductance component cannot be neglected in the inductance calculations because of the relatively long end winding coils of the ADR-BLDC machine. Due to the influence of the stator on the end winding magnetic field as shown in Fig. 5.15b, the end winding leakage inductance has to be determined by 3-D analysis.

By using the determined inductance components, the total phase inductance L_{p} can be estimated by

$$\begin{aligned} L_{\mathrm{p}} &= L_{\mathrm{s}} + L_{\mathrm{g}} + L_{\mathrm{e}} \\ &\approx 82\,\mu\mathrm{H} + 115\,\mu\mathrm{H} + 16.4\,\mu\mathrm{H} = 213.4\,\mu\mathrm{H}. \end{aligned}$$ (5.12)

In addition to the self-inductance, the mutual inductances between phases are needed. However, their calculation requires more detailed analyses, like the magnetic equivalent circuit method.

The inductance analysis mentioned above shows that the analytical calculations are useful to relate the inductance values to the machine design parameters, but they can be applied, if the magnetic field problem is simplified. On the other hand, the FEM analysis is a powerful tool to calculate the inductances, if accurate results are required, since machine design, nonlinear material characteristics, influence of the phase currents, PM flux and magnetic interactions between the phases can be considered. If stator and rotor are axially aligned, 2-D FEM analysis model is appropriate in order to calculate the phase inductances except for the end winding inductance L_e, which must be analyzed by 3-D FEM analysis.

If the ferromagnetic materials are assumed to have linear magnetic characteristics, the self- and mutual inductances are independent of the phase currents and the PM flux. However, the stator and rotor cores have non-linear magnetic characteristics. Therefore, the effects of the axial rotor position dependent PM flux and the phase currents on the magnetic field distribution have to be considered.

The influence of the phase current characteristics on the inductances of the ADR-BLDC machine is analyzed in detail in [80]. It is reported that the phase current waveforms are critical in determining the rotor position dependency of the self- and mutual inductances, i.e. their waveform deviates a lot from ideal. Therefore, it is proposed to calculate the phase inductances by assuming ideal phase current waveforms in case of unknown phase current characteristics.

Consequently, first the axial rotor position dependent phase flux linkages are calculated with ideal phase currents by using 2-D FEM analysis. Then, L_k and M_{k-l} are determined from these results according to Eqn. 3.14 and Eqn. 3.15 as follows:

$$L_k(\theta_m) = \frac{\Delta \Psi_k(\theta_m)}{\Delta i_k(\theta_m)} \text{ at } \Delta i_k = 2\,\text{A and } \Delta i_{\text{others}} = 0\,\text{A} \tag{5.13}$$

$$M_{k-l}(\theta_m) = \frac{\Delta \Psi_k(\theta_m)}{\Delta i_l(\theta_m)} \text{ at } \Delta i_l = 2\,\text{A and } \Delta i_{\text{others}} = 0\,\text{A} \tag{5.14}$$

Fig. 5.16 shows the self- and mutual inductances of phase 1 over an electrical period at no-load. The self-inductance has higher values, if the conductors of the phase stand over the pole-gap ($\theta_m = 0°$ and $22.5°$), and lower values, if the conductors of the phase stand over the middle of the PM ($\theta_m = 11.25°$ and $33.75°$). This is a result of the magnetic field distribution in the stator. As shown in Fig. 4.5, the flux density is higher in the teeth that stay in the pole middle at no-load. Therefore, the inductance is lower in the pole middle due to the relatively low permeability of the stator teeth. The mutual inductances depend on the rotor angular position as well as on the relative position of the related phase conductors. The mutual inductance between two stator phases is lower, if the closest conductors of the related phases stand over different magnet poles. The self-inductance is always positive. The mutual inductances are positive or negative depending on the assigned current direction of the closest coils of the related phases.

The effects of the phase currents can be seen in Fig. 5.17, which shows the inductances at 300 Nm. As can be seen, the rotor position dependency of the inductances changes under load and the phase currents affect the self-inductance more than the mutual inductances. Moreover, the ratio between the self- and mutual inductances, phase coupling k, depends highly on the phase currents.

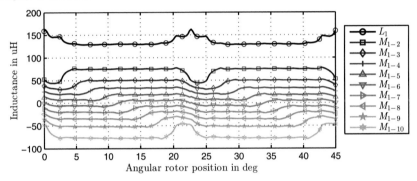

Figure 5.16: Simulated self- and mutual inductances with open circuited stator windings by 2-D FEM analysis

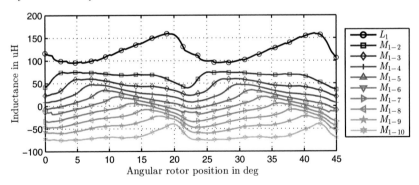

Figure 5.17: Simulated self- and mutual inductances at $T = 300$ Nm by 2-D FEM analysis

The end winding inductance component is not considered so far in the above presented results. Therefore, this inductance component is separately analyzed in the following text, where the self- and mutual end winding inductances are determined from the 3-D FEM analysis results by using the method proposed in [81]. According to this method, first the phase inductances are calculated with the 3-D models having different axial stator core lengths without changing the end winding length and design. Then the inductance of the stator core and the end winding regions are determined from these results.

The end winding self- and mutual inductances are calculated at no-load for the following simplified machine geometries:

1. *without* stator core and *without* rotor

2. *with* stator core and *without* rotor

3. *with* stator core at constant relative permeability and *without* rotor

The results of case 1 and 2 are listed in table 5.2. As can be seen, the end winding inductance shows great differences. This confirms that the end winding leakage flux is influenced by the stator core, thus the end winding inductance calculations must be done in the presence of the stator core.

Case	L_1 (µH)	M_{1-2} (µH)	M_{1-3} (µH)	M_{1-4} (µH)	M_{1-5} (µH)	M_{1-6} (µH)	M_{1-7} (µH)	M_{1-8} (µH)	M_{1-9} (µH)	M_{1-10} (µH)
1	7.99	5.54	3.69	2.27	1.08	0.00	-1.08	-2.27	-3.69	-5.54
2	23.24	13.05	8.66	5.36	2.55	0.00	-2.55	-5.36	-8.66	-13.05

Table 5.2: End winding self- and mutual inductances of phase 1 calculated for case 1 and 2 by 3-D FEM analysis

In both cases, the influence of the PM flux on the stator core magnetic field distribution is not considered. Therefore, the saturation of the stator core due to the PM flux is represented in case 3 by changing the relative permeability of the stator core. The rotor is excluded from the 3-D model, in order to reduce the computation time. This model simplification is applicable, since the influence of the rotor core on the flux distribution in the end winding region can be neglected because of the large effective air-gap. In the 3-D FEM model, the relative permeability of the stator core in the z direction $\mu^*_{r,z}$ is set to 10 based on the results given in Fig. 4.4, and the relative permeability of the stator core in the x-y plane $\mu^*_{r,xy}$ is varied.

The inductances calculated for case 3 are given in table 5.3. As the relative permeability of the stator core decreases, the end winding inductance decreases as well. This shows that the end winding inductances vary with the saturation of the stator. If the stator core is absolutely saturated, the inductances will be the same as the results *without* the stator. However, the difference between the extreme values is relatively small compared to the inductances of the stator core region. According to the 2-D FEM analysis results, the relative permeability of the stator core is in the range of $200 - 300$. Therefore, the end winding inductance values determined for case 3 at $\mu^*_{r,xy} = 300$ are used in the base speed range.

$\mu^*_{r,xy}$	L_1 (μH)	M_{1-2} (μH)	M_{1-3} (μH)	M_{1-4} (μH)	M_{1-5} (μH)	M_{1-6} (μH)	M_{1-7} (μH)	M_{1-8} (μH)	M_{1-9} (μH)	M_{1-10} (μH)
1000	25.35	12.92	8.31	5.28	2.40	0.00	-2.40	-5.28	-8.31	-12.92
300	20.55	12.57	8.31	5.14	2.45	0.00	-2.45	-5.14	-8.31	-12.57
200	19.67	12.36	8.18	5.05	2.40	0.00	-2.40	-5.05	-8.18	-12.36
100	17.74	11.75	7.84	4.84	2.30	0.00	-2.30	-4.84	-7.84	-11.75
50	15.35	10.70	7.24	4.48	2.13	0.00	-2.13	-4.48	-7.24	-10.70
25	12.82	9.24	6.34	3.95	1.88	0.00	-1.88	-3.95	-6.34	-9.24

Table 5.3: End winding self- and mutual inductances of phase 1 calculated for case 3 by 3-D FEM analysis

5.3.2 Mechanical Field Weakening Range

The no-load inductance values are compared in Fig. 5.18 under two extreme conditions, i.e. if the stator/rotor are axially aligned and the rotor is completely removed. From these results it can be concluded that an increase in the self- and mutual inductances is expected as the axial stator/rotor overlap length decreases. Consequently, the inductance of the ADR-BLDC machine is expected to vary with the axial rotor position; however, an extreme change is not probable. On the other hand, the influence of the additional magnetic flux components due to the rotor/stator misalignment is ignored in this comparison. Therefore, the effects of these additional magnetic field components on the phase inductances are analyzed in this section. Based on these results, a possible model reduction method is proposed.

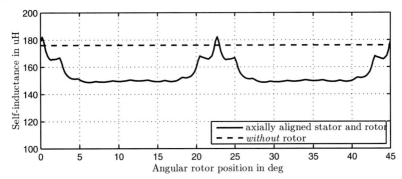

Figure 5.18: Simulated phase inductances including the end winding inductances from table 5.2 and table 5.3

The computing time required for the inductance calculations with 3-D FEM analysis may be critical with limited computational power. The reason is that the discretization of the calculation volume is crucial to be able to calculate accurate inductance values, since the flux linkages have to be determined accurately in case of small changes in the phase currents. Therefore, another method is needed especially to determine the dependency of the inductances on the phase currents at different axial rotor positions. If the electric machine can be divided into axially homogeneous regions, the inductances of these regions can be determined by using 2-D FEM analysis. By considering only the mechanical structure of the ADR-BLDC machine in the mechanical field weakening range, three axially homogeneous regions in the positive z plane can be defined, as shown in Fig. 5.19:

- inner stator region $(0\,\mathrm{mm} < l_z < (70\,\mathrm{mm} - l_{s-r})/2 < l_z)$,

- outer stator region $((70\,\mathrm{mm} - l_{s-r})/2 < l_z < 35\,\mathrm{mm})$,

- end winding region $(35\,\mathrm{mm} < l_z < 55\,\mathrm{mm})$.

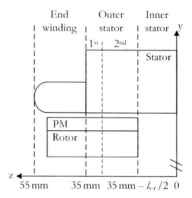

Figure 5.19: Regions with axially symmetric construction in the mechanical field weakening range

However, the magnetic flux distributions of these regions are not axially homogeneous. This can be seen in Fig. 5.20, where the relative permeability of the stator teeth is illustrated. The inner stator region has higher local relative permeability values than the outer stator region. Additionally, the permeability of the stator teeth is smallest at the outer end of the stator due to the influence of the overhang magnetic flux.

In the first place, the inductance of the inner stator region is examined by using the 2-D FEM model *without* rotor. The self-inductance values are calculated by setting a constant relative permeability for the stator core material. These results are shown Fig. 5.21. In this case, the self-inductance does not depend on the phase currents because of linearized material characteristics. Moreover, the inductance value changes with high relative permeability values. However, this change remains insignificant. The difference in the inductance values is for example less than 1%, if $\mu_r > 2000$, from which the μ_r values in the inner stator region are higher, as shown in Fig. 5.20. In addition to these calculations, the inductances of the inner stator region are calculated under load conditions

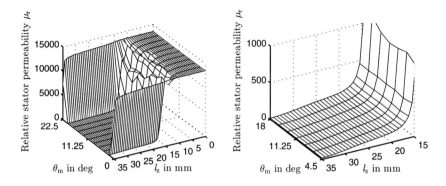

(a) $0\,\mathrm{mm} < l_z < 35\,\mathrm{mm}$ and $0\,^\circ < \theta_r < 22.5\,^\circ$ **(b)** $15\,\mathrm{mm} < l_z < 35\,\mathrm{mm}$ and $4.5\,^\circ < \theta_r < 18\,^\circ$

Figure 5.20: Relative permeability of the stator teeth at r=r4+3 mm (Fig. 4.6) depending on the angular rotor position at $l_{s-r} = 40\,\mathrm{mm}$

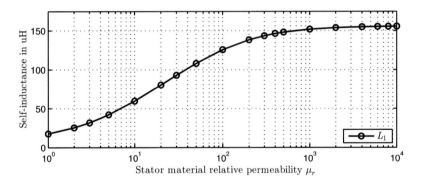

Figure 5.21: Self-inductance calculated at varying constant relative stator material permeabilities by using the 2-D FEM model *without* rotor

with non-linear stator core chracteristics. It is found out that the phase currents slightly affect the inductances in the absence of the PM rotor. Accordingly, it is concluded that the flux created by the phase currents is not high enough to saturate the stator core. In the light of these results, the following assumptions can be made:

- The influence of the inner fringing effects on the phase inductance can be neglected.

- The phase inductance of the inner stator region does not vary with the angular rotor position.

- The specific phase inductance of the inner stator region is assumed to be constant.

- The 2-D FEM model *without* rotor can be used to estimate the inductance of the inner stator region.

As a result, the self- and mutual inductances of the inner stator region can be calculated by

$$L_{k\,\text{inner}}(l_{\text{s-r}}) = \frac{L_{k\,\text{without rotor}} \times (70\,\text{mm} - l_{\text{s-r}})}{70\,\text{mm}}, \tag{5.15}$$

$$M_{k-l\,\text{inner}}(l_{\text{s-r}}) = \frac{M_{k-l\,\text{without rotor}} \times (70\,\text{mm} - l_{\text{s-r}})}{70\,\text{mm}}. \tag{5.16}$$

In order to calculate the inductances of the outer stator region, the influence of the rotor on the magnetic field distribution must be considered. As shown in section 5.1, the air-gap flux density in the outer stator region is almost constant and does not depend on the axial rotor position. Accordingly, the magnetic field distribution in the rotor between $(70\,\text{mm} - l_{\text{s-r}})/2 < l_{\text{z}} < 35\,\text{mm}$ can be assumed to be axially homogeneous. On the other hand, the stator magnetic field distribution is inhomogeneous especially in the stator teeth. The influence of this inhomogeneity on the inductance values must be analyzed.

The 3-D FEM flux density distribution results in Fig. 5.22 show that the magnetic flux distribution in the stator teeth at the outer ends of the stator is relatively homogeneous in the radial direction. This is because of the additional magnetic flux from the rotor overhang entering the stator core on its end plane. This region is called the 1st part (Fig. 5.19) and it axially extends approximately to $30\,\text{mm} < l_{\text{z}} < 35\,\text{mm}$, where the relative permeability of the stator teeth varies between 25 and 50. This condition can be approximated by fixing the relative permeability of the stator teeth in the 2-D FEM model.

Fig. 5.23 shows the inductances calculated by fixing the relative permeability of the stator teeth in addition to the result calculated with non-linear material characteristics. In case of constant stator teeth permeability, the phase self-inductance varies with rotor position because of the magnetic saturation in the rotor yoke. The results with non-linear characteristics and constant stator teeth permeability values are comparable in the pole-gap regions around $\theta_{\text{m}} = 0°$ and 22.5°. However, they differ from each other at the other rotor positions. Even if the fixed relative permeability of the stator teeth is low, the inductances calculated by assuming constant relative permeabilities are higher than the original inductance values. This is due to fact that the saturation of the stator, especially the stator teeth, is not uniform but depends on the radial position, if the characteristic of the material is represented by a B-H curve. As a result, the flux due to the phase currents concentrates on the regions with higher permeability, causing lower inductances, since the MMFs of the phase coils are not interlinked as good as in the case with constant relative stator teeth permeability.

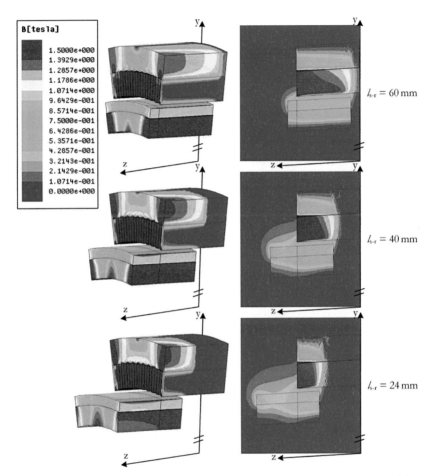

Figure 5.22: Magnetic flux density distribution on the surface of the 3-D FEM model and in the y-z plane at l_{s-r} =60 mm, l_{s-r} =40 mm and l_{s-r} =24 mm

As can be seen in Fig. 5.23, the inductance values calculated by using this approach at μ_r=50 and μ_r=25 have comparable magnitudes with the original inductance waveform. However, these inductances under load are expected to deviate from each other. Therefore, the method with constant stator teeth relative permeability is appropriate for calculating the inductances of the 1st part of the outer stator region.

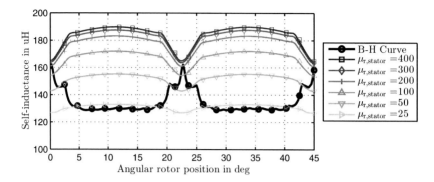

Figure 5.23: Phase inductances calculated at varying constant stator teeth permeabilities by using the 2-D FEM model *with* rotor

The inductances of the 2nd part of the outer stator region are analyzed by using the method proposed in [82]. This method is used for the numerical analysis of IPM rotor brushless machines with a fixed rotor overhang, where the remanent magnetization B_r of the PM material in the 2-D FEM model is increased, in order to consider the flux due to the fixed rotor overhang. In order to achieve this, first B_r is determined as a function of the fixed rotor overhang length, and then a 2-D FEM model is used to calculate the machine parameters. This method is valid for IPM rotor PM brushless machines, since the air-gap flux increases, if the rotor is axially longer than the stator [82]. However, it cannot be directly used for SMPM rotor brushless machines, since the rotor overhang flux barely contributes to the air-gap flux (Fig. 5.3). Furthermore, a higher PM flux causes saturation of the rotor yoke, affecting the phase inductances especially, if phase coils are located over the pole-gaps. Therefore, the inductance results calculated with this method are not valid, if the phase coils are in the pole gap regions. Nevertheless, it is an effective method to examine the influence of the saturation of the stator material. Fig. 5.24 shows the inductances calculated by using the 2-D FEM model at different B_r values over an electrical period. Additionally, the maximum magnetic flux density values in the stator teeth $B_{teeth,max}$ are listed in table 5.4. These results show that the phase inductance highly depends on the saturation of the stator core.

B_r in T	0.8	0.9	1.0	1.1	1.24	1.4	1.5	1.6	1.7	1.8
$B_{teeth,max}$ in T	0.98	1.10	1.21	1.32	1.45	1.57	1.63	1.66	1.69	1.71

Table 5.4: Maximum value of magnetic flux density in the stator teeth $B_{th,max}$ at varying constant PM material remanent magnetizations B_r

The 2nd part of the outer stator region is divided into small regions, and the maximum flux density of these regions is determined. Afterwards, the appropriate B_r values are assigned to each region according to its maximum flux density. Since a B_r value lower

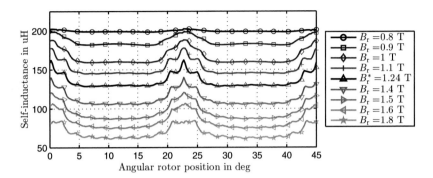

Figure 5.24: Phase inductances calculated at varying constant PM material remanent magnetizations B_r, where $B_r=1.24$ T is the original value, by using the 2-D FEM model *with* rotor

than the original value is assigned to the regions close to the inner edge and a higher B_r value to the regions close to the outer edge, the resultant estimation error for the original 2-D FEM model is under 10%. Therefore, the inductances of the 2nd part of the stator outer region can be estimated by using the original 2-D FEM model.

The lacking inductance part in the mechanical field weakening range is the end winding inductance. The influence of the rotor yoke on the flux generated by the end winding currents can also be neglected here due to the large effective air-gap. The end winding inductance is expected to decrease, as the rotor is axially displaced because of the saturation of the stator core end regions. The axial flux components in the stator core due to the rotor overhang stay almost constant for l_{s-r} values shorter than 50 mm. Consequently, the end winding inductances are expected to be constant at short l_{s-r}. Since the magnetic flux distribution depends on the rotor position, the end winding inductance is expected to depend also on the rotor position. However, this position dependency is neglected due to the relatively small end winding inductance values compared to the stator core inductance. By using the results given in table 5.2 and table 5.3, the end winding inductance depending on the axial stator/rotor overlap length is estimated as follows:

- at rotor base position: $L_e(l_{s-r} = 70\,\text{mm}) \doteq 20\,\mu\text{H}$,

- with axially displaced rotor: $L_e(l_{s-r} > 50\,\text{mm}) \doteq 17.5\,\mu\text{H}$ & $L_e(l_{s-r} \leq 50\,\text{mm}) \doteq 15\,\mu\text{H}$,

- *without* rotor: $L_e = 23.2\,\mu\text{H}$.

The proposed model reduction method for the inductance calculations and the used FEM models are summarized in Fig. 5.25. Fig. 5.26 and Fig. 5.27 show the estimated self-inductance L_1 and mutual inductance M_{1-2} values depending on the axial rotor position.

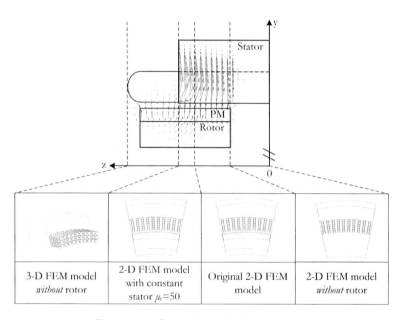

Figure 5.25: Proposed model reduction method

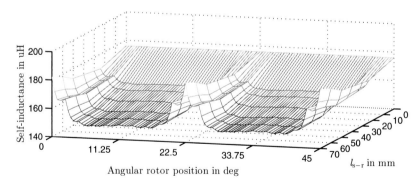

Figure 5.26: Estimated self-inductance L_1 determined by using the model reduction method, where the values *without* rotor are shown at $l_{\mathrm{s-r}} = 0\,\mathrm{mm}$

As a result, the self-inductance values increase as the rotor is axially displaced, whereas the mutual inductance values slightly decrease. Therefore, the magnetic couplings be-

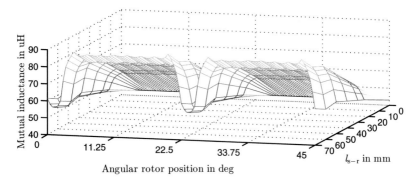

Figure 5.27: Estimated mutual inductance M_{1-2} determined by using the model reduction method, where the values *without* rotor are shown at $l_{s-r} = 0\,\mathrm{mm}$

tween phases are lower in the mechanical field weakening range. The dependency of the inductances on the axial rotor displacement is modeled in the dynamic simulation model developed in chapter 6 by using the calculated inductance values in this section.

5.4 Measurements and Validation

In the previous sections, the back EMF waveforms, torque constants and phase inductances are calculated by using FEM analysis. In these calculations, some physical effects, such as the influence of the induced eddy current field on the magnetic field distribution, are neglected. Furthermore, three-dimensional magnetic field problems are approximated by using two-dimensional analysis, as in the model reduction method proposed for the inductance calculations. Although these simplifications have been justified, the most reliable method to verify the calculation results is the comparison with experimental results. In literature, there exist no detailed analysis results for ADR-BLDC machines, thus the validation of the calculation results is especially important in this case due to the new features of the analyzed ADR-BLDC machine.

In this section, first the measurement setup used for the back EMF measurements is introduced. Then the experimental test results for the back EMF waveforms are shown, and these are used to validate the 3-D FEM analysis results. Finally, the no-load phase inductances estimated by the proposed model reduction method are validated by using the measured inductance values. The torque production of the ADR-BLDC machine is examined in this chapter by assuming ideal phase current waveforms, which cannot be realized in the measurement setups. Therefore, the torque produced with the real phase current characteristics and the phase inductances under load are analyzed and validated in the next chapter.

5.4.1 Drag Test Measurement Setup

Open circuit drag tests are performed to measure the back EMF characteristics and the no-load losses of the prototype ADR-BLDC machine. These tests are carried out on an existing drag test measurement setup. This measurement setup is chosen because of two reasons: firstly, the axial rotor position can easily be adjusted and secondly, the losses of the test machine can directly be determined.

A photo of the test bench is shown in Fig. 5.28. In this setup, an induction machine is used as drive machine, which is coupled to the shaft of the test machine through a gear belt, and the rotational speed of the test machine is set by controlling this drive machine. The machine under test is composed of housing, stator and rotor. Fig. 5.29 shows the machine parts that are used in the back EMF measurements. There are other stator and rotor constructions that are used in no-load loss measurements, which are introduced later in section 7.3.

In this test setup, the housing of the ADR-BLDC machine, which is shown in Fig. 5.29a, has a supporting function the rotor shaft being supported by the bearings mounted in the housing. The openings left around the housing are the places for the integrated power electronic modules. The stator in Fig. 5.29b is called the stator *with* winding, in order to differentiate it from the stator *without* winding, which is used in the loss measurements.

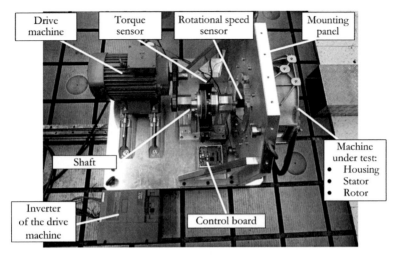

Figure 5.28: Photo of the drag test setup

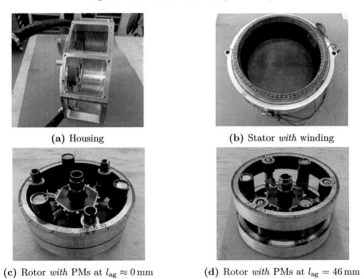

(a) Housing

(b) Stator *with* winding

(c) Rotor *with* PMs at $l_{ag} \approx 0\,mm$

(d) Rotor *with* PMs at $l_{ag} = 46\,mm$

Figure 5.29: Parts of the machine under test used in the back EMF measurements

The stator *with* winding is composed of the stator core, winding, temperature sensors, Hall effect sensors and stator support parts. Fig. 5.29c and Fig. 5.29d show the rotor with PMs with the axial gap between the rotor cylinders being zero $l_{ag} \approx 0$ mm (l_{s-r}=70 mm) and maximum $l_{ag} = 46$ mm (l_{s-r}=24 mm), respectively. The axially symmetric parts of this rotor and the axial displacement mechanism can also be seen in these photos. The axial rotor displacement mechanism is deactivated, and the axial rotor position is mechanically fixed to avoid any changes in the axial rotor position during the measurements.

Open circuit drag tests are performed under the following operational conditions:

- 6 different axial stator/rotor overlap lengths; $l_{s-r} = 70$ mm, 60 mm, 50 mm, 40 mm, 30 mm and 24 mm,

- varying rotational speeds up to 9000 rpm,

- stator temperatures $20°C - 30°C$, $50°C - 60°C$, and $80°C - 90°C$.

The measured maximum rotational speed is limited in the back EMF measurements, in order to avoid any isolation failure in the stator winding. Accordingly, the maximum voltage of a phase winding is restricted to 400 V. Moreover, the rotor speed is limited to 9000 rpm due to the mechanical strength of the prototype rotor shafts.

5.4.2 Back EMF

The back EMF waveforms measured and simulated at 1000 rpm are shown in Fig. 5.30 over an electrical period. The markers and the solid curve represent the measurement and 3-D time stepping FEM results, respectively. The number of measurement points is reduced to ease a visual comparison. This figure shows that the simulation results are consistent with the test bench results. Consequently, the previously discussed effects of the axial stator flux and the end winding geometry on the back EMF characteristics are experimentally verified.

The dependency of the back EMF waveforms on the rotational speed is examined by comparing the back EMF waveforms at low and high rotational speeds. The back EMF waveforms for 250 rpm and 9000 rpm at $l_{s-r} = 24$ mm are normalized with the rotational speed of the operation as

$$e_{norm}(\theta_m, \omega_m) = \frac{e(\theta_m, \omega_m)}{\omega_m}. \tag{5.17}$$

The normalized measurement results are compared in Fig. 5.31. These waveforms are nearly identical, and the small differences can be explained by the rotational speed fluctuations during the measurements. As a result of this comparison, the effect of the eddy current reaction field in the stator due to the time-varying axial flux components can be ignored in the rotational speed range of 0 rpm $< n < 9000$ rpm.

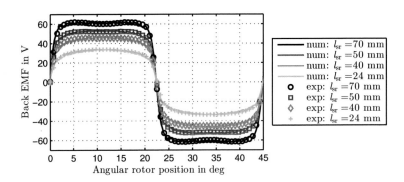

Figure 5.30: Validation of the back EMF waveforms at 1000 rpm and stator temperatures between $20°C - 30°C$

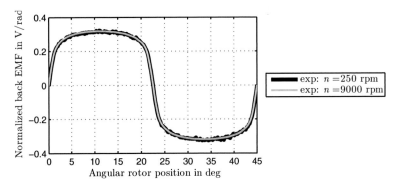

Figure 5.31: Comparison of the back EMF waveforms for 250 rpm and 9000 rpm at $l_{s-r} = 24\,\text{mm}$ and stator temperatures between $20°C - 30°C$

By using Eqn. 3.10, the amplitude of the phase flux linkage $\hat{\Psi}_{PM}$ can be calculated from the measured back EMF characteristics as:

$$\hat{\Psi}_{PM}(\theta_r, \omega_r, l_{s-r}) = \frac{1}{2} \int_{\theta_{r0}}^{\theta_{r0}+\pi} \left| \frac{e(\theta_r, \omega_r, l_{s-r})}{\omega_r} \right| d\theta_r . \tag{5.18}$$

The amplitudes of the flux linkage calculated from the measured back EMF results are shown in Fig. 5.32 at varying l_{s-r} as a function of the rotational speed at stator temperatures between a) $20°C - 30°C$ and b) $80°C - 90°C$. In both of these diagrams, the amplitudes of phase flux linkages at a given axial rotor position are constant for the measured rotational speeds. This confirms the previous discussion about the induced eddy current reaction field. The amplitude of the flux linkages decreases as the stator temperature increases, which is caused mainly by the temperature dependent characteristics of the PMs. Although the rotor temperature is not recorded during the measurements, an increase in the stator temperature implies also an increase in the rotor temperature. The difference in $\hat{\Psi}_{PM}$ between the measurements at $20°C - 30°C$ and $80°C - 90°C$ is about 4%.

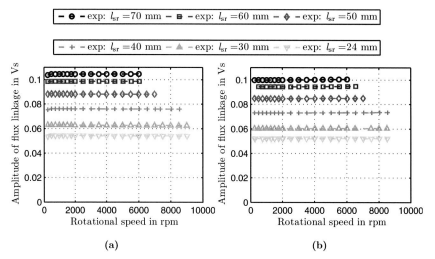

(a)   (b)

Figure 5.32: Amplitude of the flux linkage calculated from the measured back EMF waveforms at stator temperatures between a) $20°C - 30°C$ and b) $80°C - 90°C$

Additionally, the average values of these calculated flux linkages are determined for varying l_{s-r} and these average values are compared with the 3-D FEM analysis results, as shown in Fig. 5.33. In the 3-D FEM model, the PMs are modeled with a remanence magnetization B_r value equal to $1.24\,T$ corresponding to $20°C$ according to [78]. The difference between the experimental results at stator temperatures between $20°C - 30°C$ and the numerical results is under 2.5%, which is a good agreement despite the complexity of the analyzed system.

5.4.3 No-Load Phase Inductance

The self- and mutual inductance values of the ADR-BLDC machine are measured at no-load (measurement current $\leq 200\,mA$) at stand still *without* and *with* rotor at varying axial rotor positions. In order to bring the rotor to a reference angular position, first the reference phase is supplied with a DC current, so the coils of this phase align in the

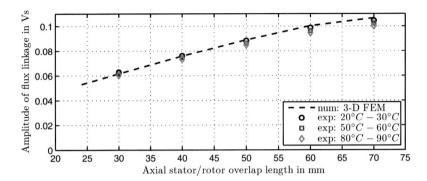

Figure 5.33: Validation of the amplitude of the flux linkage calculated by Eqn. 5.18

pole-gaps. Then the rotor is fixed at this position, in order avoid any oscillations due to the supplied measurement current. These measurements are performed with an LCR-Bridge (appendix A) at 20 rotor positions over an angular span of a pole pair.

The calculated and measured inductance values of a phase in the base speed range are illustrated in Fig. 5.34 with lines and markers, respectively. The calculated phase inductance is determined as the sum of the inductance of the stator core region calculated by 2-D FEM model (Fig. 5.16) and the end winding region calculated by 3-D FEM model (table 5.3). The numerical results for the self-inductance values are in very good agreement with the measured inductance values. Moreover, the calculated mutual inductance values have comparable characteristics with the measured ones, except that they are higher in the pole middle. However, this difference is not significant and less than $15\,\mu H$.

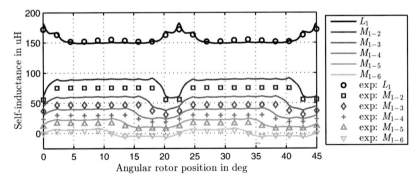

Figure 5.34: Validation of the self- and mutual inductance values at no-load in the base speed range

Fig. 5.35 shows the measured and estimated self-inductance values depending on l_{s-r}, where the results without rotor are shown at $l_{s-r} = 0$ mm. These results show that the change in the inductance values with respect to the axial rotor position is estimated well with the proposed model reduction method. However, there are small deviations between the experimental and numerical results, for example the rotor position dependency is not assessed well at low l_{s-r} lengths. Possible reasons for the differences between the measured and calculated results are the assumptions done in the model reduction process and the non-modeled properties of the ADR-BLDC machine such as the non-linear material properties.

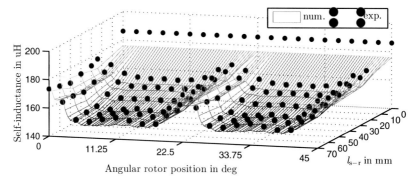

Figure 5.35: Validation of the self- inductance values at no-load in the mechanical field weakening range

The comparisons in this section show the validity of the calculated no-load self- and mutual inductance values both in the base speed and mechanical field weakening ranges.

6 Modeling and Control

In this chapter, first the developed dynamic simulation model of the ADR-BLDC drive system is introduced, and then the operational characteristics including the limits of the drive system are analyzed by using this model. Main motivation of this analysis is to examine the behavior of the drive system both in the base speed and mechanical field weakening ranges.

6.1 Dynamic Simulation Model

Electric machines can be modeled by different approaches depending on the application. At the beginning of the design process, analytical models are favorable for machine dimensioning, material selection and determination of basic machine characteristics. On the other hand, models based on numerical calculations are commonly used for optimization purposes and detailed analyses. Since an existing electric machine is analyzed in this study, the FEM analysis models introduced in section 4.1 are used to examine the characteristics of the ADR-BLDC machine. The dynamic behavior of the drive system can be simulated by coupling these FEM analysis models directly or indirectly with circuit simulations as discussed in [83] and [84]. Despite long simulation durations in case of many operating points with limited computational capacity, this approach is applicable with 2-D transient FEM analysis. However, simulation durations are not acceptable for circuit simulations coupled with 3-D transient FEM analysis. Therefore, a dynamic simulation model in the mechanical field weakening range is needed.

In [85], a 3-phase BLDC drive system including an inverter, a mechanical load, and a mathematical model of a 3-phase BLDC machine, in which real back EMF waveforms are implemented and commutation effects are considered, is modeled in the MATLAB/Simulink environment. A more detailed 3-phase BLDC drive system model is proposed in [86], which is similarly developed in MATLAB/Simulink and Plecs. In this second model, parameters of the BLDC machine including back EMF waveforms, inductances and cogging torque values are determined by FEM analysis neglecting the magnetic saturation effects and then implemented in the dynamic simulation model. In [87], non-linear magnetic material characteristics are also considered in the FEM model, and the angular rotor position dependency of the phase inductance values is implemented using look-up tables in the MATLAB/Simulink environment. In a similar way, a single-phase BLDC motor drive system including an inverter and a mechanical load is implemented in the

MATLAB/Simulink environment in [88]. The armature reaction is also taken into account in this model. As can be seen, it is common practice to determine the electric machine parameters by FEM analysis and to implement these parameters in dynamic simulation models. This simulation method combines the accuracy of FEM analysis with fast dynamic simulations. It is important to note that the accuracy of these models highly depends on the chosen equivalent circuit diagram and the implemented machine parameters.

Consequently, the ADR-BLDC machine drive is modeled with a dynamic simulation model that includes a phase variable model of the electric machine due to following reasons:

- This modeling approach is applicable both in the base speed range and in the mechanical field weakening range, provided that the machine parameters are known.

- Fast dynamic simulations are possible with good accuracy.

The dynamic simulation model is implemented in Ansys Simplorer v10. As can be seen in Fig. 6.1, it consists of five main parts:

- Phase variable model of the ADR-BLDC machine

- DC voltage supply

- Power electronic components

- Mechanical load

- Control block

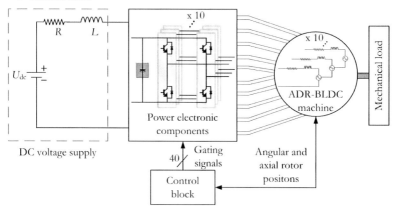

Figure 6.1: Dynamic simulation model

As shown in Fig. 6.1, the DC voltage supply with the DC cable connections is modeled by an ideal DC voltage source with a resistance and a line inductance. The DC-link capacitor and 10 full-bridge inverters are modeled in the power electronic block by using ideal IGBT and diode models from the component library of Simplorer. The gating signals of the semiconductor switches in the power electronics block are determined in the control block. Finally, the mechanical load represents the load torque and the moment of inertia of the total system.

The phase variable model of the ADR-BLDC machine is implemented based on the basic BLDC equations derived in section 3.4. Accordingly, the phases are modeled with their resistances, self-inductances, mutual inductances to other stator phases and voltage sources, which represent the back EMF waveforms, as shown in Fig. 6.2. The phase resistance is modeled by using the measured DC resistance value at room temperature, which is equal to $86.6\,\mathrm{m\Omega}$. Its frequency dependency is neglected based on the AC resistance calculations in section 7.1.1, and the effect of temperature is considered by using Eqn. 7.4. The self- and mutual inductance look-up tables are formed using the results from section 5.3. Effects of the magnetic saturation are considered in these results as a function of the DC input current I_{dc} due to the linear dependency between the amplitude of the ideal phase current and I_{dc}. Since phase coupling coefficients are easier to handle, the phase couplings that are calculated by using Eqn. 3.16 are used in the model instead of the mutual inductances. Consequently, the self-inductances and the coupling coefficients are represented in the dynamic simulation model as a function of the angular rotor position θ_r, the DC input current I_{dc}, and the axial stator/rotor overlap length l_{s-r}. The back EMF waveforms from section 5.1 are implemented in the simulation model by using look-up tables. The effect of the armature reaction on the back EMF waveform is ignored, and the back EMF of a phase is modeled as a function of the rotor rotational speed ω_r, the angular rotor position θ_r, and the axial stator/rotor overlap length l_{s-r}. Since the influence of the axial rotor displacement is considered, this electric machine model is valid in the total operation region.

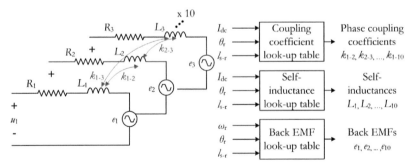

Figure 6.2: Phase variable model of the ADR-BLDC machine

6.2 Operational Characteristics of the ADR-BLDC Machine

The objective of this section is to examine the influence of the axial rotor position on the steady-state operational characteristics of the ADR-BLDC machine and to determine the operational limits of the drive system in motor mode. In these analyses, the gating signals of the semiconductor switches are generated by using the pulse amplitude modulation (PAM) method, which is explained in the following subsection.

6.2.1 Pulse Amplitude Modulation Control

In a typical PAM, the DC-link voltage U_{dc} is regulated by an additional DC/DC converter, and the electrical commutation is achieved by a multi-leg inverter [89]. Fig. 6.3 shows the block diagram of a typical PAM control with a DC/DC converter. As can be seen, only the DC input current is controlled with feedback control, thus only one controller and one current measurement is needed.

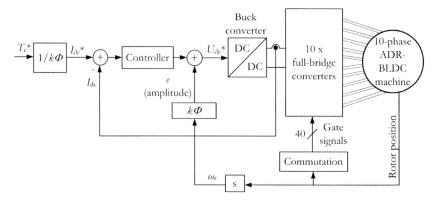

Figure 6.3: Block diagram of PAM control with DC/DC converter

The DC/DC converter can be eliminated by using the full-bridge inverters as voltage choppers in the base speed range. In this case, all full-bridge inverters are controlled with the same control signals during the phase conduction durations to regulate the phase voltages by using the PWM method. In the mechanical field weakening range, the induced back EMF of the BLDC machine is regulated by varying the axial rotor displacement, and the full-bridge inverters are only used for the electronic commutation. Both of these PAM methods

- PAM with DC/DC converter and

- PAM by using the full-bridge inverters as buck DC/DC converter

are applied to the ADR-BLDC machine. The buck type DC/DC converter is not implemented in the simulation model, but its effect is simulated by changing the input DC voltage. Since the phase currents are not directly controlled, they can highly deviate from the idealized waveform not only because of the commutation of the currents but also because of the magnetic couplings between the stator phases [90]. Therefore, phase currents highly depend on the equivalent circuit parameters making this control strategy suitable for validating the calculated machine parameters with the measurement results.

Commutation

The commutation timings have an important influence on the performance of BLDC machines. The terminology used for the commutation is illustrated in Fig. 6.4, which is valid for all control strategies examined in this study. In this figure, the voltage and current waveforms of a phase simulated by using PAM with DC/DC converter control are shown. θ_{on}, θ_{fw} and θ_{off} represent the turn-on, free-wheeling and turn-off angles, respectively. These commutation angles are defined with respect to the zero crossing point of the induced back EMF of the corresponding phase. Beginning with θ_{on} to θ_{fw}, the phase is activated. Therefore, θ_{on}-θ_{fw} is the phase conduction angle. A free-wheeling period at the end of the phase conduction between θ_{fw}-θ_{off} is included in order to limit the rate of decrease in the phase currents. This is advantageous since the interactions between the stator phases arising due to the magnetic couplings are limited in this way. The switching states of the corresponding full-bridge inverter for positive phase currents are shown in Fig. 6.5 for continuous phase currents during the phase activation interval.

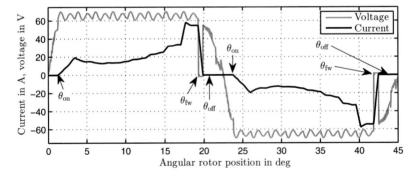

Figure 6.4: Phase voltage and current at $n = 1000\,\text{rpm}$ and $T = 100\,\text{Nm}$ with commutation angles $\theta_{\text{on}} = 5\%$, $\theta_{\text{fw}} = 85\%$ and $\theta_{\text{off}} = 90\%$ simulated by applying PAM control with DC/DC converter control

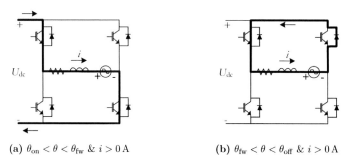

(a) $\theta_{\mathrm{on}} < \theta < \theta_{\mathrm{fw}}$ & $i > 0\,\mathrm{A}$ (b) $\theta_{\mathrm{fw}} < \theta < \theta_{\mathrm{off}}$ & $i > 0\,\mathrm{A}$

Figure 6.5: Switching states of full-bridge inverters

The commutation angles are defined as percentages of half an electrical period. For example, the commutation angles in Fig. 6.4 are $\theta_{\mathrm{on}} = 5\%$, $\theta_{\mathrm{fw}} = 85\%$ and $\theta_{\mathrm{off}} = 90\%$, which means that the phase conduction duration is set to 80% of half the electrical period. The stator phases are electrically isolated, thus there are no constraints on phase conduction durations. In order to achieve maximum torque per ampere, stator phases need to be activated when their back EMFs are almost constant. According to the back EMF characteristics of the ADR-BLDC machine, the ideal commutation angles can be defined as $\theta_{\mathrm{on}} = 10\%$ and $\theta_{\mathrm{off}} = 90\%$. However, the commutation angles must be defined according to the machine characteristics and the applied control strategy. For example, θ_{on} is set to be smaller than the ideal turn-on angle, so that the phase currents can reach higher values at the beginning of the conduction duration. Similarly, θ_{off} cannot be set higher than a certain value, since the phase current has to be decayed to zero, before the next commutation half period starts.

Influence of Magnetic Phase Couplings

Fig. 6.4 shows that the phase current waveform highly deviates from the ideal block form with PAM control. In [90], reasons for this special current waveform of a multi-phase BLDC machine with an overlapping winding configuration are examined, and it is reported that mutual inductances are responsible for this kind of current characteristics. In Fig. 6.6, all phase currents are illustrated, and the influence of the magnetic phase couplings can be observed. To begin with, the effects between the phases are explained by taking phase 5 as the reference phase. At the beginning of the conduction period, the current of phase 5 increases until phase 4 is turned on. As the current of phase 4 starts to increase, a positive voltage is induced in phase 5 due to positive magnetic coupling between these phases, and this additionally induced voltage in phase 5 limits the increase in the current of this phase. At the end of the conduction period of phase 5, there is a current peak. This peak is caused, because phase 6 is turned off. A negative voltage is induced in phase 5 over the coupling to phase 6, thus its current increases rapidly. Interactions between phases do not happen only between neighboring phases. All phase currents are affected, when a phase is activated or deactivated. Nevertheless, the phase current characteristics are highly dominated by the influence of the mutual inductances between the neighboring phases. Phase currents of a phase simulated under the same conditions with and without phase couplings are shown in Fig. 6.7. These results clearly show the effect of magnetic couplings.

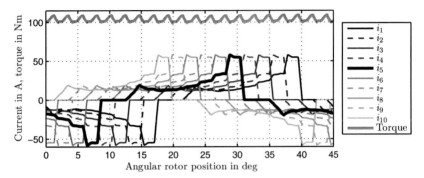

Figure 6.6: Phase current waveforms at $n = 1000\,\mathrm{rpm}$ and $T = 100\,\mathrm{Nm}$ simulated by applying PAM control with DC/DC converter

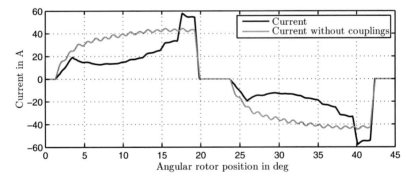

Figure 6.7: Current of a phase with and without phase couplings at $1000\,\mathrm{rpm}$ and at the same DC-link voltage, as in Fig. 6.6

It is important to note that these effects are not observed in 3-phase electric machines controlled with 3-leg inverters, even if the magnetic couplings between the phases are high. In this case, phase voltage equations can be written for a 3-phase machine as in Eqn. 6.1, if the self- and mutual inductance terms are assumed to be constant or sinusoidal with respect to angular rotor position. Moreover, the inductance matrix in this equation can be reduced to a diagonal matrix, as in Eqn. 6.2, by introducing the current constraint $i_a + i_b + i_c = 0$. This shows that the coupling terms in the inductance matrix can be mathematically eliminated.

$$
\begin{bmatrix} u_a \\ u_b \\ u_c \end{bmatrix} = R_{\mathrm{p}} \begin{bmatrix} i_a \\ i_b \\ i_c \end{bmatrix} + \begin{bmatrix} L & M & M \\ M & L & M \\ M & M & L \end{bmatrix} \frac{\mathrm{d}}{\mathrm{d}t} \begin{bmatrix} i_a \\ i_b \\ i_c \end{bmatrix} + \begin{bmatrix} e_a \\ e_b \\ e_c \end{bmatrix} \tag{6.1}
$$

$$
\begin{bmatrix} u_a \\ u_b \\ u_c \end{bmatrix} = R_{\mathrm{p}} \begin{bmatrix} i_a \\ i_b \\ i_c \end{bmatrix} + \begin{bmatrix} L-M & 0 & 0 \\ 0 & L-M & 0 \\ 0 & 0 & L-M \end{bmatrix} \frac{\mathrm{d}}{\mathrm{d}t} \begin{bmatrix} i_a \\ i_b \\ i_c \end{bmatrix} + \begin{bmatrix} e_a \\ e_b \\ e_c \end{bmatrix} \tag{6.2}
$$

Fig. 6.6 shows that the peak-to-peak torque ripple is around 10 Nm. Despite the irregular characteristics of the phase currents, the torque ripple is relatively small due to the high number of stator phases. This is a known advantage of multi-phase electric machines [91]. This torque characteristic is simulated with PAM control by assuming that the input DC voltage is regulated by a DC/DC converter. Therefore, there is a minimum number of switchings per electrical period which results in a lower torque ripple compared to the operation without a DC/DC converter [30].

The irregular phase current characteristic is not desired. First, the produced torque is proportional to the average phase current, but the copper losses are proportional to the square of the rms value. Therefore, the irregular phase currents result in higher copper losses per produced average torque. Second, the current peaks at the end of each phase conduction period result in higher current stress of the power electronic components. Accordingly, for a better system efficiency, the phase currents should be controlled separately. On the other hand, the PAM control strategy is simple, the input DC voltage is well utilized, and, as indicated before, it is suitable for the validation of the developed simulation model.

6.2.2 Operational Limits

By using the developed simulation model, it is examined whether the drive system meets the required characteristics, i.e. the specified maximum power and maximum torque requirements. Since the DC input voltage is an important limiting factor, a nominal DC voltage equal to 300 V is used and limitations due to the specifications of the power electronic components are not considered in these analyses.

Base Speed Range

First, the operational characteristics in the base speed range are analyzed by using the above defined commutation angles. The phase current and torque waveforms are shown at average torque values around 100 Nm and 200 Nm for different rotational speeds in Fig. 6.8. These diagrams show the simulation results at $l_{s-r} = 70$ mm. The phase current characteristics change with the rotational speed on the one hand due to the fact that the increase in the phase current is limited by the time constant and on the other hand due to the different U_{dc} values.

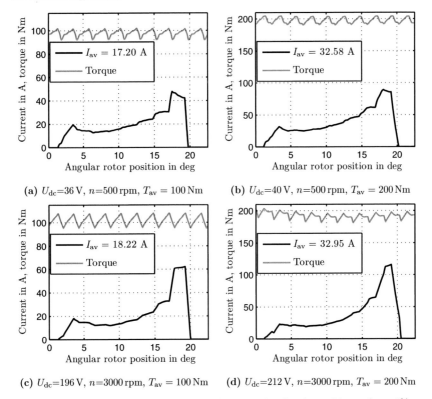

(a) U_{dc}=36 V, n=500 rpm, $T_{av} = 100$ Nm   (b) U_{dc}=40 V, n=500 rpm, $T_{av} = 200$ Nm

(c) U_{dc}=196 V, n=3000 rpm, $T_{av} = 100$ Nm   (d) U_{dc}=212 V, n=3000 rpm, $T_{av} = 200$ Nm

Figure 6.8: Phase current and torque waveforms simulated at $l_{s-r} = 70$ mm, $\theta_{on} = 5\%$, $\theta_{fw} = 85\%$ and $\theta_{off} = 90\%$ by applying PAM control with U_{dc} being regulated

Fig. 6.9 shows the phase current and torque characteristics with the full-bridge inverters being used as voltage choppers at $l_{s-r} = 70$ mm. These simulations are performed with a PWM switching frequency $f_s = 20\,\mathrm{kHz}$. Comparing the phase current characteristics shown in Fig. 6.8a and Fig. 6.8c with those in 6.9a and 6.9b, it can be concluded that the current waveforms are essentially the same except for pulsations due to switching effects. The switchings cause also higher torque ripple. These results are simulated with a synchronous PWM switching. The torque ripple can be reduced by an asynchronous switching strategy. If the DC/DC converter in Fig. 6.3 is removed and the power electronic converters are used as voltage choppers, the current and torque characteristics differ from the previously shown characteristics only if the voltage limit is not reached. Accordingly, once the rotor parts are axially displaced, both control strategies are the same. Therefore, the characteristics without DC/DC converter are only analyzed in the base speed range.

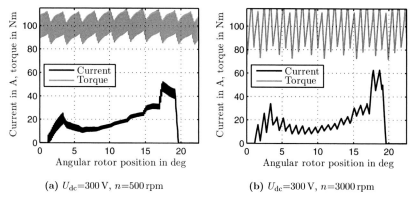

(a) U_{dc}=300 V, n=500 rpm   (b) U_{dc}=300 V, n=3000 rpm

Figure 6.9: Phase current and torque waveforms simulated at $T_{av} = 100\,\mathrm{Nm}$, $l_{s-r} = 70\,\mathrm{mm}$, $\theta_{on} = 5\%$, $\theta_{fw} = 85\%$ and $\theta_{off} = 90\%$ by applying PAM control with full-bridge inverters being used as voltage choppers

Mechanical Field Weakening

The maximum achievable torque at U_{dc}=300 V is calculated with the same commutation angles $\theta_{on} = 5\%$, $\theta_{fw} = 85\%$ and $\theta_{off} = 90\%$ as a function of the axial stator/rotor overlap length l_{s-r}. These results and the characteristic line of the ADR-BLDC machine are shown in Fig. 6.10. Accordingly, the rotor parts have to be axially displaced beginning at around 4300 rpm and, at maximum rotor displacement, the rotational speed of the BLDC machine reaches a maximum value of around 10500 rpm. The limit lines have an arc shape indicating that higher rotational speeds are achievable at a lower produced torque at a given l_{s-r} and they show the speed limits of the mechanical field weakening method with a limited axial displacement range (section 2.2.3). The amplitude of the induced back EMF has to be further decreased to be able to establish higher phase currents which result in a higher average torque. As the rotor parts are axially displaced, it is harder to achieve higher torque values, since the machine torque constant decreases. Therefore, the phases have to be supplied with higher currents to produce the same output torque compared to the axially aligned stator and rotor case. In order to reach high torque values, the difference between the back EMF and U_{dc} must be high. Consequently, the mechanical field weakening range is restricted due to these facts, and the maximum rotational speed is not achievable within the range of the axial rotor displacement at U_{dc}=300 V.

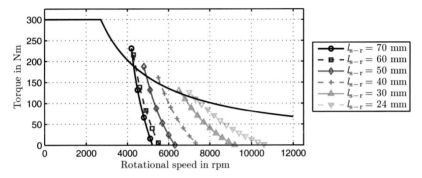

Figure 6.10: Limits of mechanical field weakening with a limited axial rotor displacement range at U_{dc}=300 V with PAM control being applied

Combined Mechanical and Electrical Field Weakening

The operational limits of the ADR-BLDC machine can alternatively be extended by using the phase advance method. Instead of decreasing the amplitude of the induced back EMF mechanically, the phase currents can be advanced by decreasing the turn-on angle θ_{on}, which can also take negative values. Even if the DC input voltage is lower than the magnitude of the back EMF, phases can be supplied with current before the back EMF takes higher values than U_{dc}. Fig. 6.11 shows the limit torque-speed line achievable with electrical field weakening. Even if a high value for the phase peak current that is equal to 150 A is assumed, only the phase advance method is not sufficient to cover the whole operating range. This is mainly due to the relatively small phase inductances of the ADR-BLDC machine, which was the motivation for implementing mechanical field weakening.

Since both the mechanical field weakening method as well as the phase advance method are not sufficient to cover the whole operating range alone, a combination of these methods must be implemented. If the phase turn-on angle is slightly advanced, the operational limits of the mechanical field weakening method also extends. If a combined mechanical and electrical field weakening method is applied, the axial rotor position and the turn-on angle of a phase have to be determined in the field weakening range. This can be achieved in different ways. The field weakening methods can for example be applied in separate

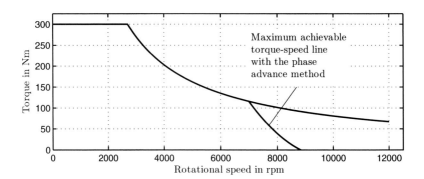

Figure 6.11: Operational limits of the drive system in the field weakening range with electrical field weakening simulated at $\hat{i}_p$ =150 A and U_{dc}=300 V by applying PAM control

operation regions, so that the maximum achievable rotational speed shown in Fig. 6.10 can be further extended by using the phase advance method. Alternatively, a combined method can be applied in the whole field weakening range. In order to optimize the field weakening strategy, losses in the field weakening range must be examined.

A combined mechanical and electrical field weakening strategy, which combines the advantage of both field weakening methods, is developed for the ADR-BLDC machine. At this point, it is assumed that the rotor parts are displaced as a function of rotational speed. The axial stator/rotor overlap length is determined as a function of rotational speed by keeping the amplitude of the induced back EMF equal to U_{dc}. As can be seen in Fig. 6.12, the rotor parts are axially displaced for $n > 5000$ rpm, and minimum l_{s-r} is reached at $n > 10000$ rpm. Since the axial rotor position is known for each rotational speed, only the phase turn-on angle is needed at each operation point. θ_{on} values in the mechanical field weakening range are calculated by using the dynamic simulation model, and Fig. 6.13 shows these new operational limits of the ADR-BLDC machine. As can be seen, the whole operating range can be covered by using this combined field weakening method.

Furthermore, the phase currents and the torque waveforms calculated in the field weakening range are shown in Fig. 6.14 at average torque values around 100 Nm and 50 Nm and at rotational speeds between 7000 rpm and 10000 rpm. The turn-on angle and the axial stator/rotor overlap length of each operation point are given in these diagrams. The previously discussed effects of the phase advancing can be seen in these results. The phase currents increase rapidly when the phase is activated due to the high voltage difference between U_{dc} and the instantaneous back EMF of the phase. After the back EMF reaches higher values than U_{dc}, the increase in the phase currents is caused due to the magnetic phase couplings. Moreover, the phase currents have better waveforms, since the peak at the end of the conduction period is lower. This is a result of the defined axial stator/rotor

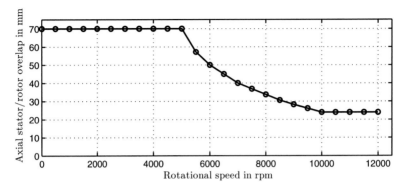

Figure 6.12: Axial stator/rotor overlap length defined as a function of rotational speed

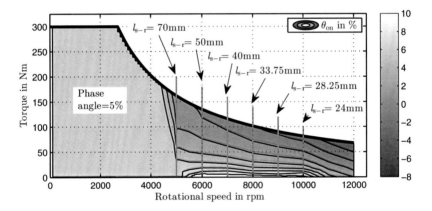

Figure 6.13: Operational limits of the drive system in the field weakening range with combined mechanical and electrical field weakening (simulated at U_{dc}=300 V by applying PAM control)

overlap length in Fig. 6.12. Accordingly, these specified l_{s-r} values are also suitable for reducing the copper losses and the stress of power electronic components.

The analyses in this section show that the total operating range cannot be covered by the mechanical field weakening at the nominal DC voltage $U_{dc} = 300$ V alone. The amplitude of the phase flux linkage can be decreased by half by reducing the axial stator/rotor overlap length from 70 mm to 24 mm. Consequently, the maximum achievable rotational speed at very low load conditions can be doubled. However, this maximum rotational

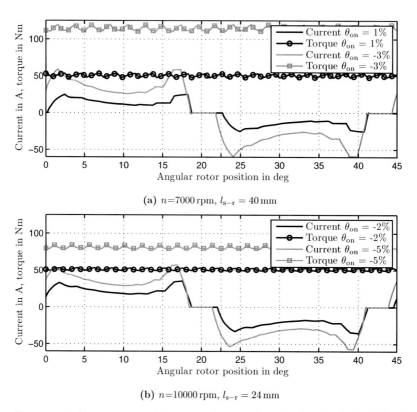

(a) n=7000 rpm, $l_{s-r} = 40$ mm

(b) n=10000 rpm, $l_{s-r} = 24$ mm

Figure 6.14: Phase current and torque waveforms with combined mechanical and electrical field weakening simulated at U_{dc}=300 V, $\theta_{fw} = 85\%$ and $\theta_{off} = 90\%$ by applying PAM control

speed decreases under high load conditions. Therefore, the speed range of the ADR-BLDC machine must be further extended by using the electrical field weakening method. The field weakening strategy, in which the mechanical field weakening and phase advance methods are combined, is a good trade-off between the additional space required to extend the mechanical field weakening range, and the higher phase inductance or lower phase flux linkage required to extend the electrical field weakening range.

6.3 Measurements and Validation

The operational characteristics of the ADR-BLDC machine with PAM control are measured in a load measurement setup. These measurements are carried out, in order to validate the developed ADR-BLDC machine model. In the following subsections, first the load test setup is introduced. Then the measured phase current characteristics are compared with the simulation results with stator and rotor being axially aligned. Finally, these comparisons are repeated for the case with an axially displaced rotor.

6.3.1 Load Test Setup

The characteristics of the ADR-BLDC machine are measured under load conditions in the load test setup with fixed axial stator/rotor overlap lengths of

1. $l_{s-r} = 70\,\text{mm}$ with axially aligned stator and rotor and
2. $l_{s-r} = 24\,\text{mm}$ with maximum mechanical field weakening.

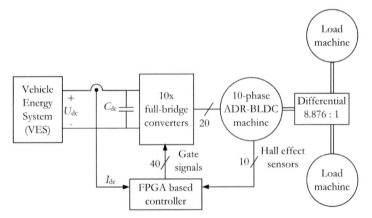

Figure 6.15: Hardware diagram of the load test setup

In these measurements, the stator *with* winding (Fig. 5.29b) and the rotor *with* PM (Fig. 5.29c and Fig. 5.29d) are used. Fig. 6.15 shows the hardware diagram of this setup. The DC input voltage U_{dc} is set by the Vehicle Energy System Unit (VES). The DC input current I_{dc} is measured by a current transducer, and the torque produced by the BLDC machine is controlled by regulating I_{dc}. The electrical commutation is achieved according to the pre-defined phase conduction durations with respect to the rotor position information from the Hall effect sensors. Two induction machines, which are connected to the ADR-BLDC machine through a differential, are used as a load. The photo of these load machines and the ADR-BLDC machine drive system is shown in Fig. 6.16. The measuring devices used in this test setup are listed in appendix A.

Figure 6.16: Photo of load test setup

The following quantities are recorded during the load tests:

- DC input voltage U_{dc} and DC input current I_{dc}
- voltages and currents of the stator phases 1, 2, 3 and 4
- torque and rotational speed of the load machines

This measurement setup is not used for the loss measurements. The reason is that only the total losses of the drive system including the differential losses can be measured in this measurement setup, since the torque and rotational speed measurements are performed on the load side.

6.3.2 Measurements with Axially Aligned Stator and Rotor

Fig. 6.17 shows the simulated current waveforms and the measurement results at four operation points. Relatively low rotational speeds around 500 rpm and rotational speeds on the limit of the mechanical field weakening range around 4600 rpm are chosen. There are small differences between the simulated and measured results corresponding to the middle of the phase conduction period. In order to find out the reasons for these differences, these simulations are repeated with the 2-D FEM analysis model coupled with circuit simulations. Since the end winding conductors cannot be modeled in the 2-D FEM model, the end

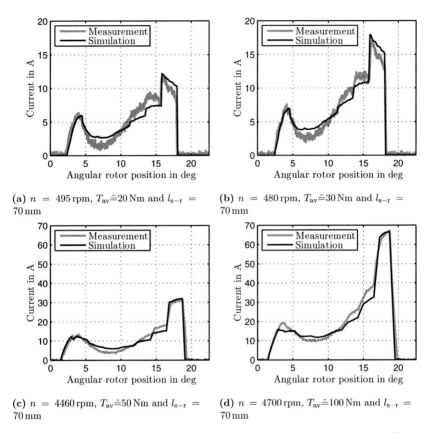

(a) $n = 495\,\mathrm{rpm}$, $T_{\mathrm{av}} \hat{=} 20\,\mathrm{Nm}$ and $l_{\mathrm{s-r}} = 70\,\mathrm{mm}$

(b) $n = 480\,\mathrm{rpm}$, $T_{\mathrm{av}} \hat{=} 30\,\mathrm{Nm}$ and $l_{\mathrm{s-r}} = 70\,\mathrm{mm}$

(c) $n = 4460\,\mathrm{rpm}$, $T_{\mathrm{av}} \hat{=} 50\,\mathrm{Nm}$ and $l_{\mathrm{s-r}} = 70\,\mathrm{mm}$

(d) $n = 4700\,\mathrm{rpm}$, $T_{\mathrm{av}} \hat{=} 100\,\mathrm{Nm}$ and $l_{\mathrm{s-r}} = 70\,\mathrm{mm}$

Figure 6.17: Comparison of the measured and simulated current characteristics with axially aligned stator and rotor at $\theta_{\mathrm{on}} = 10\%$, $\theta_{\mathrm{fw}} = 80\%$ and $\theta_{\mathrm{off}} = 90\%$ with PAM control, if U_{dc} is regulated

winding inductances are added to the circuit simulation part. The circuit simulations coupled with the 2-D FEM model deliver results very close to the developed dynamic simulation model. Consequently, it can be concluded that these differences originate from the 2-D FEM analysis model, which is used for the determination of the machine parameters in the base speed range. Possible reasons can be un-considered end effects in the 2-D FEM model, differences in material characteristics and construction tolerances that are not represented in the FEM model. Except for this, the dynamic simulation model represents very well the main characteristics of the BLDC machine, such as the influence of the mutual inductances, in the base speed range.

6.3.3 Measurements with Axially Displaced Rotor

Similar measurements are performed at the minimum l_{s-r} that is equal to 24 mm. Fig. 6.18 shows these measurement results with the simulation results at rotational speeds around 1000 rpm and 6000 rpm. The measured and simulated phase currents have also very similar waveforms in case of an axially displacement of the rotor.

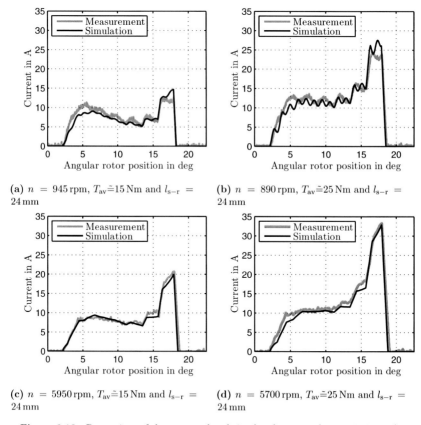

(a) $n = 945$ rpm, $T_{av} \hat{=} 15$ Nm and $l_{s-r} = 24$ mm

(b) $n = 890$ rpm, $T_{av} \hat{=} 25$ Nm and $l_{s-r} = 24$ mm

(c) $n = 5950$ rpm, $T_{av} \hat{=} 15$ Nm and $l_{s-r} = 24$ mm

(d) $n = 5700$ rpm, $T_{av} \hat{=} 25$ Nm and $l_{s-r} = 24$ mm

Figure 6.18: Comparison of the measured and simulated current characteristics with axially displaced rotor at $\theta_{on} = 10\%$, $\theta_{fw} = 80\%$ and $\theta_{off} = 90\%$ with PAM control, if U_{dc} is regulated

Additionally, the measured and simulated average torque values are compared in table 6.1. The simulated average torque values are higher than the measured ones, since the losses are not considered in the simulation models. The power difference between these results are given in the last column of this table. As can be seen, these differences are very high for such low load cases. The reasons are: First of all, the no-load losses of the ADR-BLDC machine are high at low l_{s-r} due to high mechanical losses and high additional losses arising due to the stator/rotor misalignment. According to Fig. 7.21, the measured overall no-load loss at 1000 rpm is around 200 W and at 6000 rpm more than 1600 W. In the second place, the estimated copper losses are 40 W at operation point a, 125 W at operation point b, 58 W at operation point c, and 138 W at operation point d. Additionally, the unknown losses in the differential are also included in these results. Therefore, the difference between the measurement and simulation results are acceptable.

Moreover, the phase currents have different characteristics when compared to the axially aligned stator and rotor case in Fig. 6.17. The phase current characteristics change in the mechanical field weakening range mainly due to the less trapezoidal shape of the back EMF (Fig. 5.6) and due to lower magnetic phase couplings (Fig. 5.27). The comparisons in this subsection show that the developed dynamic simulation model represents the ADR-BLDC machine well also in case of an axial displacement of the rotor.

	U_{dc}	n	Measured T_{av}	Simulated T_{av}	Difference
a	34.1 V	945 rpm	15 Nm	17.4 Nm	237.5 W
b	34.1 V	890 rpm	25 Nm	29.0 Nm	372.8 W
c	204.2 V	5950 rpm	15 Nm	19.4 Nm	2741.6 W
d	204.2 V	5700 rpm	25 Nm	29.2 Nm	2507.0 W

Table 6.1: Comparison of the measured and simulated average torques at the operating points in Fig. 6.18

6.4 Conclusion

The operational characteristics of the ADR-BLDC machine are simulated by the developed dynamic simulation model applying PAM control. One of the main objectives of this analysis was to validate the developed ADR-BLDC machine model. The modeling is complicated once rotor parts are axially displaced, since the 2-D FEM model is not valid. Since 3-D FEM analysis can only be used to calculate machine parameters at a reduced number of operating points with a limited computational power and time, a dynamic simulation model is developed, in which the effects of the axial rotor displacement on the machine parameters are considered. In this model, parameters of the ADR-BLDC machine from chapter 5 are implemented. The validation results both with axially aligned stator and rotor and with axially displaced rotor show good agreement. Consequently, it can be concluded that the developed dynamic simulation model enables very fast dynamic numerical calculations with high accuracy. Moreover, the analyses in this chapter verify the correctness of the determined machine parameters from chapter 5.

By using the developed model, the operational characteristics of the ADR-BLDC machine are analyzed in the complete operating range. These results indicate that a PAM control is not suitable due to irregular phase current characteristics, and the phase currents have to be controlled separately. Moreover, the operating limits of the drive system are analyzed, and possible field weakening strategies are discussed. Consequently, a combined electrical and mechanical field weakening strategy is proposed. By using this field weakening strategy, the total operating range is covered, and peaks in the phase current waveforms in case of simple PAM control are minimized.

Another important design criterion are the losses in the ADR-BLDC machine. Once the rotor is axially displaced, the flux in the axial center of the stator diminishes. Therefore, the core losses in the inner part of the stator can be assumed to be negligibly small at no-load conditions. Additionally, the rotor losses caused by stator spatial harmonics are expected to decrease with an increasing axial rotor displacement. On the other hand, the axial flux in the stator causes additional core losses as explained in [72]. Moreover, the time-varying magnetic flux in the end winding and conductive construction parts induces eddy currents, thus causing additional losses in these parts. Loss analysis is of great importance, in order to find out the feasibility of mechanical field weakening. Therefore, the losses in the ADR-BLDC machine are systematically analyzed in the following chapter.

7 Systematic Analysis of Losses

Loss analysis is an important part of the design of electric machines due to the desired high system efficiency and the limited cooling capacity. Therefore, the identification of losses is especially critical for automobile traction applications because of the construction space limitations. In literature, there are many publications about the determination and minimization of PM brushless machine losses. However, losses of PM electric machines with different axial rotor and stator lengths or axially misaligned stator and rotor are examined just in few studies, as e.g. in [72].

The objectives of this chapter are to determine the loss mechanisms of PM electric machines in the mechanical field weakening range and to analyze the dependency of these losses on the axial rotor position. First of all, the losses in the ADR-BLDC machine are briefly summarized for the base speed range. Then, the influence of the axial rotor displacement on the losses is discussed. Afterwards, the additional losses in the mechanical field weakening range are examined in detail. Finally, the calculated no-load losses are validated and partly extended based on the measurement results.

7.1 Base Speed Range

The ADR-BLDC machine has similar loss mechanisms as a conventional BLDC machine in the base speed range. These can be categorized as follows:

1. copper losses,

2. core losses,

3. PM losses,

4. mechanical losses.

These loss mechanisms are introduced, and calculation methods for these losses are briefly discussed in the following subsections.

7.1.1 Copper Losses

A current in a winding causes resistive losses. If a phase is supplied with a DC current, the power dissipation of this phase due to resistive losses $P_{\mathrm{cu,p}}$ can be calculated by using

$$P_{\mathrm{cu,p}} = R_{\mathrm{dc}} I^2, \tag{7.1}$$

where R_{dc} is the DC resistance of the phase that can be calculated by

$$R_{\mathrm{dc}} = \rho_{\mathrm{cu}} \frac{l_{\mathrm{cu}}}{A} = 73.3 \,\mathrm{m\Omega}. \tag{7.2}$$

In this equation, ρ_{cu} is the resistivity of copper at $20°C$, l_{cu} is the average length of the phase coil and A is the copper cross-sectional area of each turn. Consequently, the total DC copper losses in an electric machine amount to

$$P_{\mathrm{cu}} = N R_{\mathrm{dc}} I^2. \tag{7.3}$$

The resistances of the phases are measured at room temperature by using a micro ohmmeter (appendix A). The results of these measurements show that all phases have close DC resistances with an average of 86.6 mΩ. The measured DC resistance is higher than the calculated one. Possible reasons for this difference are a) a longer conductor length than the average phase coil length in Litz wire due to the transposition, b) a lower cross-section of the copper than the total of the original conductors due to the pressing operation during the production and c) the additional terminal connection resistances. Therefore, the measured phase resistance value is used in the following analysis. Moreover, the resistivity of copper increases with temperature. The dependency of the DC resistance on winding temperature can be determined by using the linear temperature coefficient of copper α_0 by

$$R_{\mathrm{dc}}(T) = R_{\mathrm{dc}}(T_0) \left[1 + \alpha_0 \left(T - T_0 \right) \right], \quad \text{where} \quad \alpha_0 = 0.00393 \,\mathrm{K}^{-1}. \tag{7.4}$$

On the other hand, the phases are supplied with alternating current. The phase resistance increases due to skin and proximity effects with the frequency of the phase currents. Therefore, AC resistance R_{ac} is defined as a function of frequency. According to Ferreira [92], skin and proximity effects are orthogonal, and the resultant copper losses can be calculated by

$$P_{\mathrm{cu}} = N \left(P_{\mathrm{cu,p,skin}} + P_{\mathrm{cu,p,prox}} \right) = N \left(R_{\mathrm{skin}} I_{\mathrm{rms}}^2 + G(\xi) \, \hat{H}_{\mathrm{e}}^2 \, l_{\mathrm{cu}} \right), \tag{7.5}$$

where

$$R_{\mathrm{skin}} = \frac{R_{\mathrm{dc}} \, \xi}{2} \frac{sinh(\xi) + sin(\xi)}{cosh(\xi) - cos(\xi)} \quad \text{for} \quad \xi = \frac{\sqrt{\pi}}{2} \frac{d}{\delta_{\mathrm{skin}}} \quad \text{and} \quad \delta_{\mathrm{skin}} = \frac{1}{\sqrt{\pi f \mu_0 \sigma_{\mathrm{cu}}}} \tag{7.6}$$

and

$$G(\xi) = \frac{\frac{\sqrt{\pi}}{2} d \xi}{\sigma_{cu}} \frac{sinh(\xi) - sin(\xi)}{cosh(\xi) + cos(\xi)}. \tag{7.7}$$

In these equations, $\hat{H}_e$ is the magnitude of the external magnetic field caused by currents in surrounding conductors, l_{cu} is the total conductor length, d is the diameter of a round conductor with an equal cross-sectional area as the rectangular conductor, δ_{skin} is the skin depth and σ_{cu} is the conductance of copper. Although the implemented Litz wire is composed of round conductors, the cross-section of the conductors is highly deformed due to pressing into a rectangular coil shape. Therefore, the effect of frequency is estimated by using the equations for square-shaped conductors. The equations for round conductors are given in [92]. R_{skin} and $G(\xi)$ depend on the winding geometry and the skin depth, thus they change with frequency. If the frequency is zero, R_{skin} is equal to R_{dc}.

Fig. 7.1a shows the ratio R_{skin}/R_{dc}, where the effect of the skin effect on the phase resistance can be seen. According to this result, the phase resistance increases less than 1% due to skin effect for frequencies under 1 MHz. This frequency is much higher than the maximum fundamental frequency of the ADR-BLDC machine (1600 Hz) and the applied switching frequency (20 kHz). Therefore, the increase in copper losses in the ADR-BLDC machine due to skin effect can be neglected.

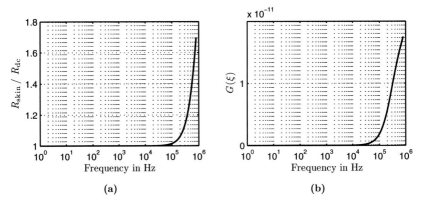

Figure 7.1: a) Ratio of R_{skin} to R_{dc} as a function of frequency and b) proximity effect coefficient $G(\xi)$ as a function of frequency

The frequency dependent proximity effect coefficient $G(\xi)$ is similarly shown in Fig. 7.1b. According to this result, this coefficient is in the range of 10^{-14} for frequencies under 0.1 MHz and in the range of 10^{-12} for frequencies between 0.1 MHz and 1 MHz. Additionally, a representative $\hat{H}_e$ value is required to determine the range of the copper losses due to the proximity effect. By using Eqn. 5.4 and Eq. 5.5, the magnetic field intensity on phase coils in a slot opening can be calculated by

$$\hat{H}_e(x) = n_c \frac{\hat{i}\, x}{b\, h}. \tag{7.8}$$

An extreme value of $\hat{H}_e$ calculated for $x = h$ and $\hat{i} = 100\,\mathrm{A}$ is equal to $265\,\mathrm{kA/m}$. When this value is used in the second term of Eqn. 7.5, the total copper losses due to the proximity effect are less than $0.2\,\mathrm{W}$ for frequencies under $0.1\,\mathrm{MHz}$ and less than $20\,\mathrm{W}$ for frequencies under $1\,\mathrm{MHz}$. Although the magnetic field intensity in the end winding region is expected to be lower, the end winding conductors are also included in l_cu. This calculation shows that the copper losses due to proximity effect are low below $1\,\mathrm{MHz}$ even for the chosen high magnetic field intensity value.

The discussions above show that the influence of skin and proximity effects is not critical for the operating frequencies of the ADR-BLDC machine. Therefore, the copper losses are calculated using

$$P_\mathrm{cu} = N R_\mathrm{dc} I_\mathrm{rms}{}^2. \tag{7.9}$$

Fig. 7.2a shows the calculated copper losses at different operating points, if the phases are supplied with ideal phase currents. Since this loss mechanism is only related with the geometry of the winding and the phase currents, it is calculated for the whole operating speed range by considering the effect of the mechanical field weakening on the machine constant. Fig. 7.2b shows the machine constant $k\phi$ of the ADR-BLDC machine as a function of $l_\mathrm{s-r}$. Since $k\phi$ decreases in the field weakening range, the phase current has to be increased as $l_\mathrm{s-r}$ decreases, provided that the produced torque is the same. Therefore, the copper losses increase in the mechanical field weakening range.

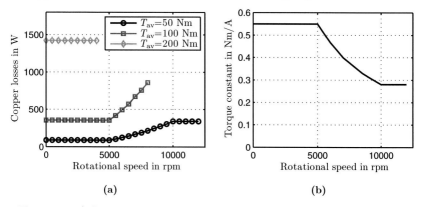

(a)   (b)

Figure 7.2: a) Copper losses calculated with ideal phase currents and b) torque constant as a function of rotational speed based on the operational limits shown in Fig. 6.13

7.1.2 Core Losses

The stator and rotor cores of the ADR-BLDC machine are exposed to a time-varying magnetic field which causes core losses in these machine parts. According to Bertotti [93], magnetic losses in soft ferromagnetic materials per unit volume P_{fe} can be decomposed into the sum of hysteresis losses P_h, classical eddy current losses P_c, and excess eddy current losses P_e, as in Eqn. 7.10. Hysteresis losses are a static loss component caused by the hysteresis characteristics of soft ferromagnetic materials. Classical and excess eddy current losses are dynamical loss components caused by eddy currents in the materials. Classical eddy current losses represent the losses caused by eddy currents induced by a time-varying magnetic field, whereas excess eddy current losses represent the losses originating from domain wall motions.

These loss components are commonly calculated in the frequency domain by

$$P_{fe} = P_h + P_c + P_e = k_h f \hat{B}^\beta + k_c \left(f\hat{B}\right)^2 + k_e \left(f\hat{B}\right)^{1.5}, \tag{7.10}$$

where the material specific loss coefficients k_h, k_c, k_e and exponent β are used [93].

Loss calculations in the time domain are more suitable for the ADR-BLDC machine due to the non-sinusoidal magnetic field characteristics. The dynamic loss components can be rewritten as a function of time derivative of the magnetic flux density. However, the calculation of the hysteresis losses in the time domain can be done based on some assumptions. In [94], a calculation method for the hysteresis loss component in the time domain is proposed. By using this method, the instantaneous value of the hysteresis losses can be determined. Accordingly, the instantaneous core loss per volume $p_{fe}(t)$ can be calculated in the time domain by

$$p_{fe}(t) = \frac{1}{\pi} k_h B_m cos(\theta)\frac{dB}{dt} + \frac{1}{2\pi^2}k_c\left(\frac{dB}{dt}\right)^2 + \frac{1}{C_e}k_e\left|\frac{dB}{dt}\right|^{1.5}, \tag{7.11}$$

in which the exponent β in Eqn. 7.10 is assumed to be equal to 2, and B_m has to be determined depending on the previous magnetization of the material. The definition of θ can be found in [95], and the constant $C_e = 8.76$ as explained in [94]. As can be seen, only the loss coefficients of the materials are needed for this instantaneous loss calculation.

Numerical calculation methods, especially FEM analysis, are used as standard tools for the approximate loss computation. The time domain calculation method explained above is implemented in the used FEM software Maxwell v14. The core losses in the stator and rotor cores are simulated by using 2-D FEM analysis under different operating conditions using the phase currents from chapter 6 with PAM control. The current waveforms are imported into the 2-D FEM analysis model. This one-directional coupling between circuit analysis and FEM analysis is an efficient tool for loss calculations. By using this method, the core losses are calculated for the total rotational speed range, and Fig. 7.3 shows these results. It is important to note that the calculation results in the base speed range

$n < 5000$ rpm are valid only for normal operation of the ADR-BLDC machine. The results at higher rotational speeds are used later in section 7.2 to estimate the losses in the mechanical field weakening range.

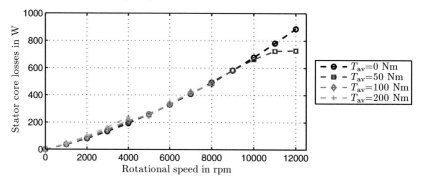

Figure 7.3: Average stator core losses simulated by 2-D FEM analysis

The calculated core losses in the rotor core are not shown due to their very low values under 5 W. According to [96], space harmonics due to the conductor distribution in slots, time harmonics in the stator currents, and stator slot harmonics are the causes of rotor losses including PM losses in PM brushless machines. Due to the large effective air-gap between the stator and rotor cores of the ADR-BLDC machine, the influence of the stator on the magnetic field distribution in the rotor is low. Therefore, the losses in the rotor yoke are much smaller than the stator core losses.

Moreover, the effects of the applied control methods can be seen in Fig. 7.3. In the base speed region, the core losses slightly increase with the produced torque. In the field weakening range, in which mechanical field weakening method as well as phase advance method are applied, losses are very close to their no-load values. After the maximum axial rotor displacement is reached at $n = 10000$ rpm, the rotational speed of the ADR-BLDC machine is only extended by the phase advance method. Therefore, for $n > 10000$ rpm, the core losses under load are lower than the no-load losses. This diagram shows that the core losses depend on the characteristics of the currents and the used control strategy.

7.1.3 Permanent Magnet Losses

PMs are hard ferromagnetic materials and have the same loss mechanisms as soft ferromagnetic materials. On the other hand, minor loops in the demagnetization curve of PMs are usually neglected. In [97], it is reported that the recoil lines of Nd-Fe-B magnets do not form a significant minor loop, unless the B axis is crossed. In [96], the findings of Fakuma et al. [98] are analyzed, and it is concluded that hysteresis losses in the PMs are generally small, unless there is an appreciable variation in the magnet flux density at high frequencies. Accordingly, hysteresis losses of the PM material are neglected and only the classical eddy current loss component is analyzed in terms of PM losses.

Due to the finite axial length of the PM magnets, the PM losses must be calculated using 3-D analysis. Only the approximated PM losses can be calculated using 2-D analysis, since only the induced eddy currents in the z direction are considered. An accurate PM loss computation is out of the scope of this study. Therefore, PM losses are estimated by 2-D FEM analysis. Fig. 7.4 shows these simulation results. As can be seen, the PM losses highly increase with the phase currents, and they are low compared to the stator core losses.

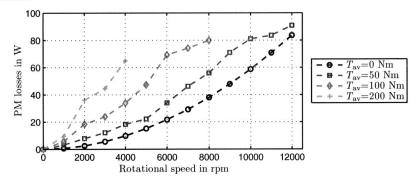

Figure 7.4: Average PM losses calculated by 2-D FEM analysis

7.1.4 Mechanical Losses

The last loss mechanism deals with mechanical losses, which are composed of windage, friction and bearing losses. Windage and friction losses depend on the construction of the electric machine and on the rotational speed. Bearing losses depend both on rotational speed and on the forces acting on the bearing as well as on the characteristics of the bearing. Mechanical losses must be determined to be able to evaluate the measurement results. Therefore, the total mechanical losses are measured at varying rotational speeds and axial rotor positions. These results are presented in section 7.3.

7.2 Field Weakening Range

The effects of the additional magnetic field components due to stator/rotor misalignment are considered in this section from the perspective of losses. An accurate loss analysis by using time stepping 3-D FEM analysis is only possible with simplified simulation models for some loss components. Therefore, the previously defined regular losses changing with the axial rotor position are briefly discussed based on the magnetic field distribution results with the axially misaligned stator and rotor. Afterwards, the new loss mechanisms arising in the field weakening range are analyzed in detail.

7.2.1 Overview

The terminology used in this subsection and an overview of the losses in the field weakening range are shown in Fig. 7.5.

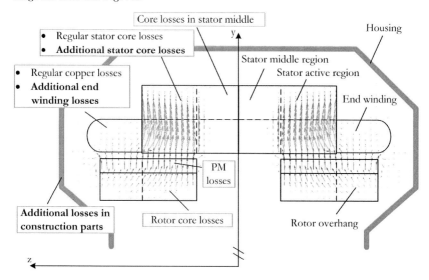

Figure 7.5: Overview of the losses in the mechanical field weakening range and the magnetic flux density distribution in the y-z plane

The axial rotor displacement does not have a remarkable influence on the regular copper losses, except that the copper losses per produced average torque increase due to the reduced torque constant, as discussed in the previous section. However, there is another loss mechanism which is caused by the time-varying PM magnetic field in the end winding regions. Eqn. 7.5 can be written for the field weakening range by considering these losses in the end winding conductors called additional end winding losses $P_{cu,fw,ew}$ as follows:

$$P_{cu,fw} = N\left(P_{cu,p,skin} + P_{cu,p,prox}\right) + P_{cu,fw,ew}. \qquad (7.12)$$

Since these additional losses are caused by an external time-varying magnetic field, they can also be classified as copper losses due to the proximity effect.

In order to determine the dependency of the core losses on the axial rotor position, the influence of the axial rotor displacement on the magnetic field distribution is analyzed. Fig. 7.6 shows the magnitude of the magnetic flux density B_{mag} on the surface of the stator core and the rotor at no-load.

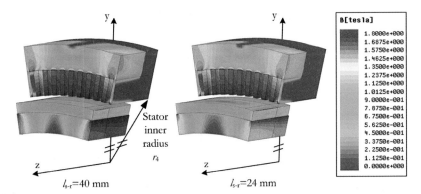

Figure 7.6: No-load distribution of the amplitude of the magnetic flux density at $l_{s-r} = 40\,\text{mm}$ and $l_{s-r} = 24\,\text{mm}$

The inhomogeneity in the stator is mainly caused by the axial components of the magnetic flux density. These axial flux components are directed in the stator in radial direction, thus the radial flux components increase as well. Therefore, it is difficult to separate the flux components from each other and consequently the losses. In order to simplify the stator core loss calculations, losses in the stator are divided into 1) regular stator core losses that are proportional to l_{s-r} and 2) additional stator core losses due to the stator/rotor misalignment $P_{fe,fw,stator,add}$. Based on this assumption, the total regular core losses in the active stator regions are estimated by

$$P_{fe,fw,stator,reg} \hat{=} \frac{P_{fe,stator}(l_{s-r} = l_z)\, l_{s-r}}{l_z} . \qquad (7.13)$$

These losses are calculated by using the results shown in Fig. 7.3. This assumption is plausible for the ADR-BLDC machine due to the fact that the air-gap radial magnetic flux density B_{rad} is almost homogeneous in the axial stator/rotor overlapping region (Fig. 5.3b).

Therefore, the rest of the stator core losses can be assumed to be caused by the stator/rotor misalignment. In these additional losses, core losses due to axial flux components and additional radial flux components in the stator core as well as losses in the stator middle region are included. Phase currents cause a time-varying magnetic flux also in the middle of the stator, which is not exposed to the PM flux except for the inner fringing effects. These losses are calculated by removing the rotor for $l_{s-r}=l_z$. As can be seen in Fig. 7.7, these losses are low compared to the stator core losses.

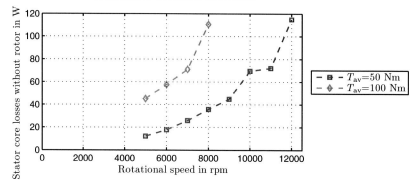

Figure 7.7: Average stator core losses *without* rotor calculated by 2-D FEM analysis

Fig. 7.5 shows that only some of the rotor overhang flux closes over the stator. Therefore, the stator slot harmonics can only cause losses in a limited part of the rotor overhang. Moreover, the effective air-gap between the stator core and the rotor overhang is larger than the nominal value of the effective air-gap, which results in lower stator time harmonics in the overhang regions. Consequently, the rotor core losses are expected to be much lower in the rotor overhang regions. As a result of these statements and the fact that the losses in the rotor represents only a small amount of the total losses, core losses in the rotor overhang regions are neglected, and the rotor core losses in the mechanical field weakening range are estimated by

$$P_{\text{fe,fw,rotor,reg}} = \frac{P_{\text{fe,rotor}}(l_{s-r} = l_z)\, l_{s-r}}{l_z}\,. \tag{7.14}$$

Similar to the rotor core losses, the PM losses in the rotor overhang regions are expected to be much lower than the PM losses in the axial stator/rotor overlapping region. Therefore, the PM losses can also be estimated to decrease proportional to l_{s-r}.

Since the magnetic flux is not only present in the stator and rotor but also in the other machine parts, the additional flux components cause losses in the construction parts. These parts are supporting, covering, ancillary and mounting components such as shaft, bearings, housing, cooling system, bolts and screws. This additional loss component is difficult to localize because of the complex geometry of the construction parts and the dependency of the magnetic field distribution on the axial stator/rotor overlap length. In Fig. 7.5, the additional losses in construction parts are only shown in the housing of the ADR-BLDC machine, since they mainly occur in this construction part as discussed in the next section.

The discussions above can be summarized as follows. Copper losses in the base speed range change in the mechanical field weakening range because of the change in the machine constant. Core and PM losses of the ADR-BLDC machine in the base speed range can be assumed to decrease linearly with the axial stator/rotor overlap length $l_{\mathrm{s-r}}$. Moreover, the core losses due to phase currents in the middle part of the stator are not critical. Additionally, three additional loss mechanisms caused due to stator/rotor misalignment are identified. These are:

1. additional end winding losses,

2. additional stator core losses,

3. additional losses in construction parts.

In the following subsections, these additional losses of the ADR-BLDC machine are determined by using the numerical results. Furthermore, the calculation methods developed to calculate them are introduced. These analyses are carried out in two steps. First, the no-load additional losses are calculated. Then the additional losses under load are estimated in the light of the additional no-load losses.

7.2.2 Additional No-load Losses

Additional End Winding Losses

The proximity effect losses in phase coils due to a time-varying external PM flux are analyzed in studies on slotless PM electric machines. In [99] and [100], armature field and PM field are superposed, and the resulting proximity losses are calculated by using the method proposed by Ferreira in [92]. In [101], it is proposed to use the equation derived for eddy current losses in a thin and long conductor (width $\ll$ length). Accordingly, specific eddy current losses of a rectangular thin conductor in the presence of an external magnetic field can be calculated by

$$\frac{P_{\mathrm{cu}}}{V} = \frac{1}{24}\,\sigma_{\mathrm{cu}}\,d^2\,\omega^2\,\hat{B}^2, \tag{7.15}$$

where V is the volume of the conductor [67]. The factor 24 is replaced by 32 for round conductors [101]. The second approach is followed to calculate the additional end winding losses.

It is possible to use Eqn. 7.15 for calculating the additional end winding additional losses, if the magnetic field distribution in the end winding regions is known. Therefore, the magnetic field distribution in the end winding regions is calculated for different axial stator/rotor overlap lengths by using 3-D FEM magnetostatic analysis. In Fig. 7.8, the radial flux density distribution in one of the end winding regions ($35\,\mathrm{mm} \le l_z \le 55\,\mathrm{mm}$) is shown for $l_{\mathrm{s-r}}{=}40\,\mathrm{mm}$ and $l_{\mathrm{s-r}}{=}24\,\mathrm{mm}$. From these results, it can be concluded that the end winding conductors are exposed to the non-sinusoidal magnetic fields which highly depend on $l_{\mathrm{s-r}}$. Moreover, the magnetic field distribution in the end winding region changes with the radial position of the end winding conductors. Since the eddy currents caused by the magnetic flux along the conductor are neglected in Eqn. 7.15 based on the assumption that width $\ll$ length, the equivalent flux density values that are orthogonal to the direction of current in the end winding conductors are used when calculating the additional end winding losses. Actually, the determination of the effective magnetic field in the end winding regions is very difficult because of the complex construction of the end winding conductors, such as twisted wires, changing layers and return structures, which are complicated to model.

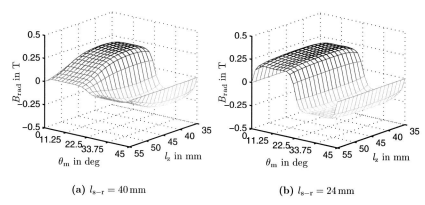

(a) $l_{\mathrm{s-r}} = 40\,\mathrm{mm}$ (b) $l_{\mathrm{s-r}} = 24\,\mathrm{mm}$

Figure 7.8: Magnetic flux density distribution in the end winding at $r = r_4 + 3\,\mathrm{mm}$, as defined in Fig. 7.6

The method used to calculate the additional end winding losses is summarized in Fig. 7.9. In the first step, the magnetic flux density values in the end winding conductors are calculated by using 3-D FEM magnetostatic analysis. In the second step, the local values of the magnetic flux density orthogonal to the current direction at discrete positions are extracted from these results. However, the time characteristics of the local flux values are needed which can be calculated with transient analyses. At this point, the transient analyses are not performed due to the required high computational time with a limited computing power. Instead, in the third step, the time characteristics of the magnetic flux density at discrete positions in a chosen radial end-winding cross-section are approximated by using the known angular position dependency of the magnetic flux density. It is sufficient to know the flux density values in a radial cross-section since the average of the specific losses in each radial cross-section is the same. In the fourth step, the harmonic coefficients of the magnetic flux density in each of the discrete positions in the chosen cross-section are determined with fast Fourier transformation (FFT). In the last step, Eqn. 7.15 is used to calculate the specific end winding losses at each of these discrete positions. In these calculations, the losses due to the field harmonics are calculated, as explained in [67], by using the harmonic coefficients. Finally, an average value of the specific losses is calculated, and the total additional end winding losses are calculated by multiplying it with the end winding conductor volume.

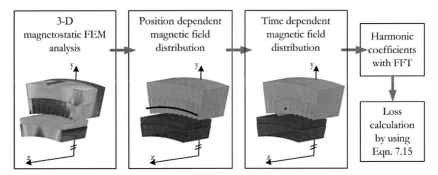

Figure 7.9: Method used to calculate the additional end winding losses

The resulting end winding losses including both end winding sides are shown in Fig. 7.10 at varying l_{s-r}. The additional end winding losses change proportional to square of rotational speed, since the skin effect is not considered. The skin effect is not taken into account in these calculations because of the high critical frequency over $180\,\mathrm{kHz}$ for copper with a diameter $d = 0.3\,\mathrm{mm}$, which is calculated by

$$f_e = \frac{4}{\pi \mu_0 \sigma d^2} \tag{7.16}$$

[72]. Moreover, these results also show that the additional end winding losses highly depend on the axial stator/rotor overlap length l_{s-r}. As l_{s-r} decreases, the additional end winding losses increase, since the rotor overhang under the end winding conductors gets longer. It is important to note that the end winding losses do not increase significantly, if l_{s-r} is decreased from $l_{s-r} = 30\,\mathrm{mm}$ to $l_{s-r} = 24\,\mathrm{mm}$. This is because the rotor overhang is slightly longer than the axial end winding length at $l_{s-r} = 24\,\mathrm{mm}$. Accordingly, the additional end winding losses are expected to increase very little once the rotor overhang is axially longer than the end winding conductors.

The analyses in this subsection show that the additional end winding losses cannot be neglected due to high losses over $800\,\mathrm{W}$ at $12000\,\mathrm{rpm}$, even if the winding is built from

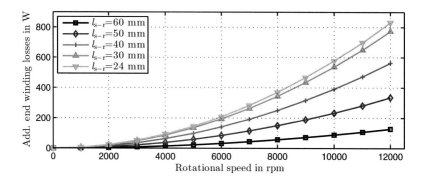

Figure 7.10: Calculated additional end winding losses

a Litz wire. This is another reason for using the Litz wire in the analyzed ADR-BLDC machine. Additional end winding eddy current losses can be reduced by decreasing the size of the insulated conductors, decreasing the pole pair number of the machine, reducing the volume of the end windings and increasing the distance between the rotor overhang and the end winding conductors.

Additional Stator Core Losses

The second additional loss mechanism of a ADR-BLDC machine is the additional stator core losses. Stator core losses do not decrease proportional to l_{s-r}, but the overhang PM flux causes additional losses in the stator core. As discussed before, to simply the loss calculations, stator core losses that decrease proportional with l_{s-r} are called regular stator core losses, whereas the remaining losses are the additional stator core losses. Due to the three-dimensional magnetic flux distribution in the stator core, these losses can be calculated, provided that the loss coefficients of the stator material in the presence of both main (radial) and axial excitation are known. These data are not available, since electrical steel is mainly used in electric machines with a nearly 2-D magnetic flux distribution. Actually, losses under 3-D magnetization conditions are subject of many publications as [102] and [103]. However, soft magnetic composites (SMC) are the main area of research, since SMCs are used in 3-D flux applications. Therefore, the influence of the axial rotor displacement on the additional stator core losses are approximated by using the results of 3-D magnetic field distribution and 2-D loss analysis at varying axial rotor positions.

Fig. 7.11 shows the radial flux density B_{rad} and axial flux density B_z distributions in the teeth and in the stator yoke for l_{s-r}=40 mm. As can be seen, the stator part that stands over the axial gap between the rotor parts 0 mm $< l_z <$ 10 mm is flux free. Accordingly, the no-load losses of this stator region are zero. In the outer stator part at

$10\,\mathrm{mm} < l_z < 35\,\mathrm{mm}$, neither B_{rad} nor B_z is axially homogeneous in the teeth. Moreover, the axial flux components due to the rotor overhang flux and the inner fringing effects merge because of the short axial length of the studied ADR-BLDC machine. The axial flux in the teeth is much higher than the axial flux in the stator yoke. The magnitude of the flux in the stator yoke increases locally due to the increase in the radial flux component in the teeth.

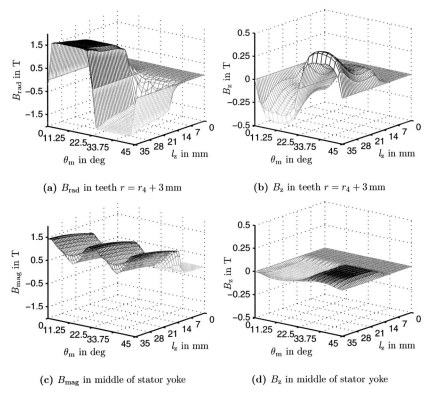

(a) B_{rad} in teeth $r = r_4 + 3\,\mathrm{mm}$

(b) B_z in teeth $r = r_4 + 3\,\mathrm{mm}$

(c) B_{mag} in middle of stator yoke

(d) B_z in middle of stator yoke

Figure 7.11: Magnetic flux density distribution in the stator teeth and in the stator yoke for $l_{\mathrm{s-r}}=40\,\mathrm{mm}$

In order to analyze the stator core losses, the magnetization properties of electrical steel in the stator are required. Even though the stator material is a non grain-oriented electrical steel, its magnetization properties are not isotropic in the rolling direction [72]. Moreover, magnetization characteristics of this material also differ between the rolling plane and the perpendicular direction to the rolling plane. Therefore, magnetic properties of the stator material are different, if the magnetic flux additionally has axial components. However, due to the fact that the axial flux components are relatively small compared to the flux in radial direction, the unknown effect of the axial flux on the magnetization properties is neglected. Consequently, the stator material is assumed to have isotropic magnetic characteristics in an electrical sheet.

The magnetic isotropy assumption has the consequence that hysteresis losses at a given frequency can be assumed to be directly proportional to $\hat{B}_{\mathrm{mag}}^{\beta}$ (Eqn. 7.10). Therefore, hysteresis losses in the stator core at different axial rotor positions are compared by using the following proportionality factor

$$P_{\mathrm{h}} \propto \frac{1}{n_{\mathrm{e}}} \sum_{i=1}^{n_{\mathrm{e}}} \hat{B}_{\mathrm{mag},i}^{\beta}, \tag{7.17}$$

where n_{e} is the total number of the evaluation points.

Furthermore, eddy current losses with misaligned stator and rotor are more complicated to predict than hysteresis losses. The paths of eddy currents induced by the flux components in the x-y plane are restricted by the width of the lamination sheets. However, induced in-plane eddy currents are only restricted by the dimensions of the BLDC machine and at relatively high frequencies due to the skin effect. The known eddy current loss coefficients, k_{c} and k_{e}, are only valid for the flux components in the x-y plane. Consequently, a possible change in the eddy current losses due to the magnetic flux density in the x-y plane B_{xy} are approximated roughly according to Eqn. 7.10 by

$$P_{\mathrm{c,xy}} + P_{\mathrm{e,xy}} \propto \frac{1}{n_{\mathrm{e}}} \sum_{i=1}^{n_{\mathrm{e}}} \hat{B}_{\mathrm{xy},i}^{2} \tag{7.18}$$

neglecting the changes in the frequency spectrum of the magnetic flux density and also overestimating the excess eddy current losses by using an exponent that is equal to 2 instead of 1.5. Since the axial flux components are much lower than the radial ones, the proportionality factor in this equation can also be approximated using B_{mag} instead of B_{xy}. This leads to an easier approximation, since both loss proportionality factors defined in Eqn. 7.17 and Eqn. 7.18 have the same axial rotor position dependency. Moreover, these equations will be the same, if the exponent in Eqn. 7.17 is assumed to be 2. This value is also assumed to be 2 in the numerical loss calculations [94]. Therefore, this assumption can be made.

The effect of the axial rotor position on the total of P_{h}, $P_{\mathrm{c,xy}}$ and $P_{\mathrm{e,xy}}$ is determined by calculating the above defined proportionality factor at discrete evaluation points in the total volume of the stator core. The resulting normalized proportionality factors are given in Table 7.1. The ideal normalized proportionality factors in this table are determined so that the regular stator core losses change directly proportional to $l_{\mathrm{s-r}}$. Accordingly, the difference between the calculated proportionality and ideal proportionality factors gives the relative value of the total of P_{h}, $P_{\mathrm{c,xy}}$ and $P_{\mathrm{e,xy}}$. The total of these additional stator core losses, which are calculated by using the factors given in Table 7.1 and the no-load stator core losses from Fig. 7.3, are shown in Fig. 7.12. As can be seen, these losses are less than 200 W at maximum rotational speed. Therefore, the estimated part of the additional stator core losses does not have a significant effect.

l_{s-r}	70 mm	60 mm	50 mm	40 mm	30 mm	24 mm
Factor	1	0.96	0.86	0.73	0.59	0.50
Factor ideal	1	0.86	0.71	0.57	0.43	0.34

Table 7.1: Normalized proportionality factor for the sum of the additional hysteresis losses and the eddy current losses due to the flux parallel to the x-y plane calculated by 3-D FEM analysis with Eqn. 7.17 and ideal factor calculated linearly decreasing with l_{s-r}

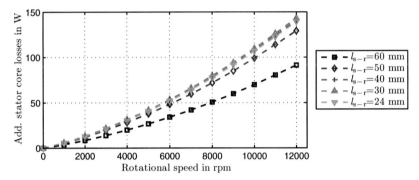

Figure 7.12: Sum of the approximated additional hysteresis and eddy current losses due to the flux in the x-y plane without load

On the other hand, eddy current losses caused by the axial flux components cannot be estimated with the known loss coefficients. Due to the complexity of this loss mechanism, an experimental method is chosen to determine the total additional stator core losses. This method and the measured additional stator core losses are presented in section 7.3.5.

Additional Losses in Construction Parts

Considering the design of the ADR-BLDC machine that is shown in Fig. 7.13, it can be concluded that the housing and the support of the stator are exposed to a time-varying magnetic field, and the losses in the other construction parts can be neglected. The mechanical displacement mechanism and the shaft are not included in this figure, since these parts rotate synchronously with the rotor, thus they are not exposed to a time-varying PM field. The stator support is used to fix the stator to the inner side of the housing, and the losses in the stator support are expected to arise in this connection part. The construction of the housing and the stator support is complex, and the used stators have different stator support constructions. On the other hand, the construction of the outer side of the housing is same in the used setups, and the stator support does not extend beyond the outer stator corner. Therefore, additional losses in construction parts are analyzed only in the outer side of the housing. In the first place, this analysis is conducted in order to quantify the losses in the construction parts. Moreover, the calculated housing losses are used to estimate the additional stator core losses in section 7.3.5. For this purpose it is sufficient to know only the housing losses in one housing side.

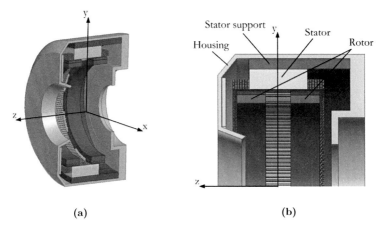

(a)   (b)

Figure 7.13: ADR-BLDC machine model including stator support and housing

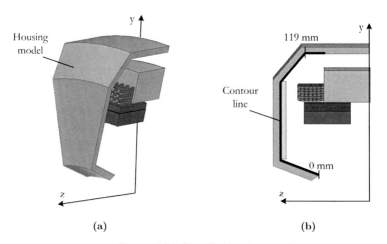

(a)   (b)

Figure 7.14: Simplified housing model

As explained in section 3.2, the housing of the BLDC machine is made of an aluminum casting alloy, which is non-ferromagnetic. Therefore, only eddy current losses are considered. In order to calculate the housing losses, the outer side of the housing is modeled in the 3-D FEM model with some geometrical simplifications. The 3-D FEM model including the simplified housing model is shown in Fig. 7.14.

The magnitude of the magnetic flux density on the inner contour line of the housing (Fig. 7.14b) is shown at varying l_{s-r} in Fig. 7.15. These results show that the magnetic flux density in the housing is locally different and highly depends on the axial rotor position. In these analyses, the mechanical displacement mechanism is not modeled based on the results of the preliminary 3-D FEM calculations, which showed that the mechanical displacement mechanism does not have a remarkable effect on the magnetic flux density distribution in the housing.

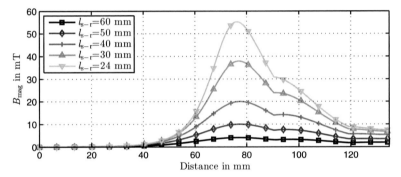

Figure 7.15: Magnitude of magnetic flux density on the inner contour line of the housing corresponding to the pole middle as shown in Fig. 7.14b

Due to the dynamic characteristics of eddy currents, magnetostatic calculations are not sufficient to determine the housing losses, and they must be calculated by using transient 3-D FEM analysis. Since the eddy current paths in the housing are not restricted, the influence of the skin effect must be considered. Therefore, the discretization of the housing must be fine enough to be able to simulate the skin effect properly, and many electrical periods have to be simulated until the magnetic field in the housing stabilizes. With a limited computational capacity, these calculations are very time consuming. Therefore, the 3-D FEM model has to be simplified. The required computational effort can be reduced by assigning constant magnetic permeabilities to the stator and rotor core materials aiming a faster convergence. Moreover, the 3-D FEM model can be simplified further by representing the effect of the stator core with appropriate boundary conditions. Concerning the computing time, the second approach is followed in this study, and the 3-D FEM model is reduced with the help of the results from the magnetostatic field analysis. This model reduction process is summarized in Fig. 7.16. In the first step, the magnetic field distribution in the housing is calculated with the original 3-D FEM model. Then the electric machine is removed from the model except for the rotor overhang. The influence of the stator is incorporated by setting proper boundary conditions on the back boundary plane of the reduced 3-D model. Finally, the boundary conditions are set separately for each axial rotor position. This reduced model, which includes the housing, the rotor overhang part and boundary conditions, is used to simulate the losses in the housing.

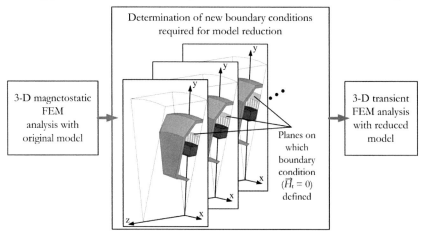

Figure 7.16: Model reduction process used to simplify housing loss calculations

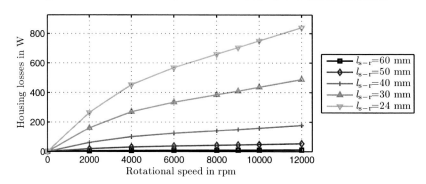

Figure 7.17: Losses simulated on one side of the housing with reduced 3-D FEM model by transient analysis

Fig. 7.17 shows the simulation results for the housing losses at varying l_{s-r}. As can be seen, the housing losses are limited by the skin depth at high speeds. Besides, they increase extremely for $l_{s-r} < 40\,\text{mm}$, as the axial distance between the housing and the outer edge of the rotor is less than $18.5\,\text{mm}$ in this case. This analysis shows that there is an average of $800\,\text{W}$ losses only on one side of the housing at minimum l_{s-r} and at maximum rotational speed.

Losses in construction parts reduce the system efficiency, hence they limit the maximum axial rotor displacement for this particular housing geometry. This type of additional losses can be reduced by proper housing design. Possible measures that can be taken to restrict the eddy currents are: a) using a housing material with lower conductance and b) using non-conductive slots in the housing or c) keeping a pre-defined distance between housing parts and the rotor overhang. Additionally, the leakage PM flux from the rotor overhang can be directed away from the housing by changes in the machine geometry, as discussed in section 8.1.3.

7.2.3 Additional Losses under Load

The magnetic field distribution and therefore the additional flux components in the mechanical field weakening range depend on the phase currents. In order to illustrate the influence of the phase currents on the additional flux components in the stator core, the magnetic field distribution of the ADR-BLDC machine is shown in Fig. 7.18 at $l_{s-r} = 40\,\text{mm}$ and $T = 170\,\text{Nm}$ ($\hat{i}_p = 50\,\text{A}$). This load condition is chosen to illustrate the influence of the phase currents in an extreme case, since $170\,\text{Nm}$ is higher than the maximum torque specification of the ADR-BLDC machine in the field weakening range. The phase currents influence the magnetic field distribution, particularly the B_z distribution in the stator core. If these results are compared with the results from Fig. 7.6 and Fig. 7.11b, it can be concluded that

- the additional flux components due to the rotor overhang slightly depend on the phase currents, since the stator teeth are highly saturated in the outer stator end regions,

- the inner fringing effects change with the phase currents and the quarter wave symmetry of B_z at a given axial position diminishes.

Consequently, the magnetic field distributions in the housing and in the end winding regions are independent of the phase currents. Therefore, the housing and additional end winding losses do not depend on the machine load, and the results shown in section 7.2.2 are also valid under load. However, an increase in the additional stator core losses is expected once the machine is loaded. This increase is expected to be uncritical, since a distinct difference in B_z is only seen at the inner side of the active stator region, where the flux density is low.

A significant part of the additional core losses are expected to be the axial flux eddy current losses [72]. Therefore, only the dependency of the axial flux eddy current losses on the phase currents is examined. The local instantaneous classical eddy current losses

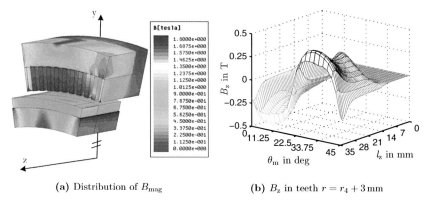

(a) Distribution of B_{mag} (b) B_z in teeth $r = r_4 + 3\,\text{mm}$

Figure 7.18: Field distribution at $l_{\text{s-r}} = 40\,\text{mm}$ and $T = 170\,\text{Nm}$

at a discrete stator position are proportional to square of the time derivative of the local magnetic flux density (Eqn. 7.11). Accordingly, proportionality factors that represent the relative stator eddy current losses due to the axial flux are calculated by

$$P_{\text{cz,av}} \propto \frac{1}{n_{\text{T}/2}} \sum_{i=1}^{n_{\text{T}/2}} (B_z(t_i) - B_z(t_{i-1}))^2 \tag{7.19}$$

over half an electrical period $T/2$ at 6 radially and 1,001 axially distributed evaluation points in a stator tooth. In this equation, $n_{\text{T}/2}$ stands for the total number of the discrete time steps over half an electrical period. This analysis is only carried out in the stator teeth due to the low B_z values in the stator yoke. The resulting average value of these factors is given in table 7.2 at $l_{\text{s-r}} = 40\,\text{mm}$ for no-load and $T_{\text{av}} = 170\,\text{Nm}$. According to these results, the axial flux losses increase by approximately 8%, if the stator phases are supplied with 50 A at $l_{\text{s-r}} = 40\,\text{mm}$. Since the eddy current losses due to axial flux can be high, their dependency on the phase currents must be considered for accurate loss calculations.

T_{av}	0 Nm	170 Nm
Factor	0.00211013	0.002272502

Table 7.2: Proportionality factors calculated by Eqn. 7.19 to determine the dependency of the axial flux eddy current losses in the stator teeth on the phase currents

As a result, the additional losses calculated at no-load give a good overall approximation, since the additional end winding and housing losses can assumed to be independent of load, and the additional stator core losses increase with load less than 8%.

7.3 Measurements and Validation of No-Load Losses

The measurement results are used to validate the calculation results presented in the previous sections and to determine the additional stator core losses. In the following subsections, first information on the measurement setup and the calculation of measurement uncertainties are provided. Afterwards, the measurement results are presented separately for each additional loss mechanism.

7.3.1 Measurement Setup

The no-load losses of the prototype electric machine are measured by performing open circuit drag tests. Detailed information about the drag test measurements is provided in section 5.4.1. In the following text, the machine parts and the machine configurations used for no-load loss measurements are introduced.

In the drag test setup, the losses of the ADR-BLDC machine are determined from torque and rotational speed measurements that are carried out directly on the electric machine side of the shaft. Therefore, this measurement setup is suitable for measuring the losses in the ADR-BLDC machine. Since only the total losses can be measured, specially constructed machine parts in addition to the presented machine parts in section 5.4.1 are used for no-load loss measurements to be able to separate the losses. Fig. 7.19 shows the photos of these additional machine parts.

(a) Stator *without* winding

(b) Single-sided PM rotor at $l_{ag} = 46\,\text{mm}$

(c) Rotor *without* PMs at $l_{ag} \approx 0\,\text{mm}$

(d) Rotor *without* PMs at $l_{ag} = 46\,\text{mm}$

Figure 7.19: Machine parts used for no-load loss measurements

The stator *without* winding in Fig. 7.19a is composed of the stator laminated steel compressed by plastic supporting parts (in green), temperature sensors built in the stator slots and stator housing. The slot openings of this stator are filled with plastic chocks (in green), in order to avoid mechanical oscillations of the stator teeth and an increase in the windage losses.

The rotor with single-sided PMs shown in Fig. 7.19b has one rotor cylinder *without* PMs on the mounting panel side and one rotor cylinder with PMs on the housing side. The rotor without PMs is shown in Fig. 7.19c and Fig. 7.19d.

The used machine configurations and their applications are listed in table 7.3.

	Stator	Rotor	Application
1	*with* winding	*without* PMs	Mechanical losses
2	*with* winding	*with* PMs	Total open circuit losses
3	*with* winding	*with* single-sided PMs	Additional end winding losses & additional stator core losses
4	*without* winding	*with* single-sided PMs	Additional end winding losses & additional stator core losses
5	-	*without* PMs	Validation of housing losses
6	-	*with* single-sided PMs	Validation of housing losses

Table 7.3: Machine configurations used for no-load loss measurements

7.3.2 Measurement Uncertainty

According to the datasheets of the used sensing devices and to the system dynamics, the absolute deviation of the torque sensor ΔT equal to 0.05 Nm and the total deviation in rotational speed measurement $\Delta\omega$ equal to 1 rad/sec are used when calculating measurement uncertainties. The measurement uncertainty in rotational speed results is not shown in the diagrams due to very small uncertainty values compared to the average value measured for rotational speed.

Since power is calculated from the corresponding torque and rotational speed values, the measurement uncertainty in power ΔP at each measurement point is calculated by using

$$\Delta P = \pm\sqrt{(T_{\mathrm{m}} \times \Delta\omega)^2 + (\omega_{\mathrm{m}} \times \Delta T)^2}. \tag{7.20}$$

In this equation, T_{m} is the average produced torque and ω_{m} is the average angular speed determined by taking the arithmetic average of the values measured over 6 ms.

All of the experimentally determined losses presented in this study are calculated by using two separate measurements except for mechanical losses. In these cases, the resulting measurement uncertainty is taken as the sum of the measurement uncertainties of each measurement:

$$\Delta P = \pm(\Delta P_1 + \Delta P_2). \tag{7.21}$$

7.3.3 Primary Measurement Results

The mechanical losses of the ADR-BLDC machine are measured first since they are needed for determining the losses in the magnetic circuit. These losses are measured with machine configuration 1 from table 7.3. Fig. 7.20 shows these results as a function of rotational speed at varying axial stator/rotor overlap lengths l_{s-r}. As can be seen in this figure, the mechanical losses are higher than 1200 W at 9000 rpm, and they slightly decrease with decreasing l_{s-r}. High mechanical losses are mainly caused by windage losses because of the complex rotor construction.

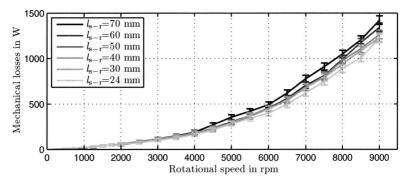

Figure 7.20: Measured mechanical losses with machine construction 1 from table 7.3

The total no-load losses in the magnetic circuit are shown in Fig. 7.21. As can be seen, these losses significantly increase in the mechanical field weakening range with decreasing l_{s-r}. These results show the high impact of the additional losses.

In Fig. 7.22, the simulated no-load losses of the BLDC machine with aligned stator and rotor are compared with the no-load losses measured in the magnetic circuit. The measurement results with configuration 2 (stator *with* winding) and with configuration 4 (stator *without* winding) are shown in this diagram. The simulated loss values are lower than the measured ones, which is often the case in loss analysis, but the difference is less than 140 W at 9000 rpm. As a result, 2-D FEM models can be used to approximate the no-load losses of the ADR-BLDC machine. Moreover, the losses measured with both stators are very close. This implies that the copper losses due to the PM flux in the slots are negligible.

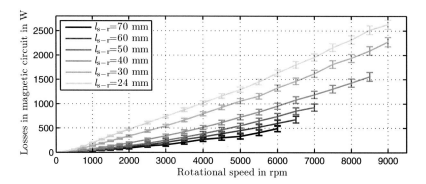

Figure 7.21: Total losses in the magnetic circuit calculated from the measurements with machine configuration 1 (mechanical losses) and machine configuration 2 (overall losses) at stator temperatures between $20°C - 30°C$

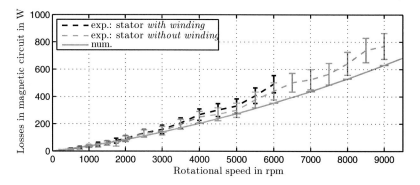

Figure 7.22: Comparison of simulated (2-D FEM) and measured no-load losses in the magnetic circuit with axially aligned stator and rotor at stator temperatures between $20°C - 30°C$

7.3.4 Validation of Additional End Winding Losses

The magnetic field distribution in the end winding regions strongly depends on the stator core. Consequently, the additional end winding losses must be measured in presence of the stator. In this study, an indirect experimental method is used to determine the additional end winding losses. In this method, first the total no-load losses are measured with the stators *with* and *without* winding (configurations 3 and 4) and then the additional end winding losses are calculated from the difference. This method is applicable due to the fact that the losses in the phase coils in the stator slots can be ignored at no-load operating conditions, since the phases are open circuited and the teeth have a much higher permeability than the material in the slots. The results from Fig. 7.22 show the validity of this assumption. Moreover, these results also show that the cores of the stators *with* and *without* winding can be assumed to have identical characteristics. The rotor with single-sided PMs is used in both machine configurations. The rotor cylinder *without* PMs is implemented at the connection side of the stator, since the supports of the used stators have different constructions at this side, and eddy current losses in these parts can result in measuring wrong additional end winding losses.

By using this indirect method, the end winding losses are determined at 3 axial stator/rotor overlap lengths, and these results are shown in Fig. 7.23. The negative loss values at low rotational speeds show that there is a systematical error which is most likely, a result of the temperature differences in the stator cores and the small construction differences between the stator supports of configurations 3 and 4. This error is taken into account in addition to the measurement uncertainty, and the total errors are indicated with the error bars. Except for some operating points, the measured and numerically determined results are consistent with each other. The possible reasons for the differences between numerical and experimental results are the effects that are not considered in numerical calculations like the complex end winding geometry and potential isolation damages within the Litz bundle or non-predictable measurement errors.

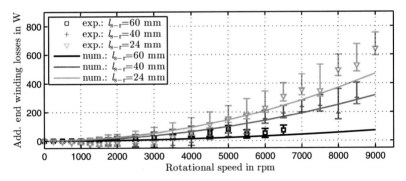

Figure 7.23: Comparison of the numerically calculated and indirectly measured additional end winding losses at stator temperatures between $20°C - 30°C$

7.3.5 Determination of Additional Stator Core Losses

Additional stator core losses of the ADR-BLDC machine are estimated by using the measurement results with configuration 3 (stator *without* winding and rotor with single-sided PMs) and the housing losses computed in section 7.2.2. The no-load losses measured with configuration 3 include regular core losses, PM losses, housing losses, mechanical losses and additional stator core losses. The rotor with single-sided PMs is used in these measurements, thus the losses in the magnetic circuit are half their original value, and there are only losses in one side of the housing. As discussed before, regular core losses and PM losses can be estimated by measurements with aligned stator and rotor. Moreover, the mechanical losses are known. In addition, the housing losses calculated in one side of the housing (Fig. 7.17) are used to determine half the additional stator core losses. The results for the total additional stator core losses are shown in Fig. 7.24. As can be seen from these curves, these losses increase proportional to square of rotational speed, and they are almost independent of the axial rotor position for l_{s-r} below 50 mm. The ADR-BLDC machine has two axially symmetrical rotor parts, thus the stator is exposed to axial magnetic flux at the inner ends of the rotor parts and at both sides of the stator. The measured additional stator core losses include all of these effects.

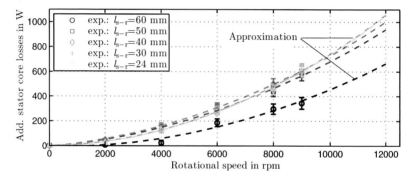

Figure 7.24: Total additional stator core losses determined using the measurement results with configuration 3 and the housing losses calculated in section 7.2.2

The total additional stator core losses are not calculated in the previous section; just the part including the hysteresis losses and the eddy current losses due to the flux parallel to the x-y plane is approximated. Consequently, the difference between the measurement results from Fig. 7.24 and the approximation results from Fig. 7.12 gives the eddy current losses caused by the axial flux. Since the approximated part of the additional stator core losses is much lower than the measured total losses, it can be concluded that the eddy current losses caused by the axial flux are the dominating loss component explaining the quadratical behavior of the additional stator core losses, which is not typical for a electrical steel with a thickness of 0.1 mm. Moreover, both results indicate that the additional stator core losses do not increase with a decreasing l_{s-r} below 50 mm. As a result, the additional flux components do not increase remarkably beyond this axial rotor position. Additionally, this shows the consistency of measurement and simulation results.

7.3.6 Validation of Housing Losses

The magnetic flux density results in Fig. 7.15 indicate that the losses in the housing arise in some specific regions. To identify these regions, a thermal image camera is used, and photos of the housing are made during a run-up at $l_{s-r} = 24\,\text{mm}$. Fig. 7.25 shows some of these photos. At the beginning of the test, the housing temperature was around $20°C$. The local warming of the housing can be seen in these photos. Because of the good heat conductance of the housing material, the local temperature differences in the housing decrease with time.

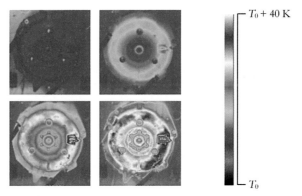

Figure 7.25: Local warming of housing measured during a run-up at $l_{s-r} = 24\,\text{mm}$

Just like additional end winding losses, housing losses can also be determined experimentally. For this purpose, additional measurements without housing or with a special housing should be carried out. Nevertheless, these measurements are not possible on the used test bench due to the supporting function of the housing. Therefore, instead of validating the simulation results for the housing losses, simplified measurements are conducted, and the simulation model is validated by using these results. In these experiments, the stator is removed, and the losses are measured at different axial rotor positions with configurations 5 and 6. Since the magnetic flux distribution is different in the absence of the stator, these results do not deliver any information about the housing losses in normal operation, but they are used to verify the 3-D FEM housing model. Fig. 7.26 shows the experimental and numerical results for the minimum axial stator/rotor overlap length. As can be seen, the housing losses increase less than proportional to rotational speed due to the skin effect. This behavior can also be seen in the numerical calculation results. The difference between experiments and numerical results is acceptable despite the simplified simulation model and the complexity of the transient skin effect analyses. These results show that the housing losses can be predicted fairly well with the developed 3-D FEM model.

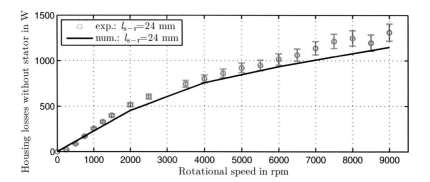

Figure 7.26: Validation of the 3-D housing model with the help of simplified measurements carried out in the absence of the stator

7.4 Summary

The no-load losses give a good overview of the additional losses in the mechanical field weakening range, since the effects of the load on the additional losses are either small or negligible. Therefore, in the first place the no-load losses are shown in Fig. 7.27.

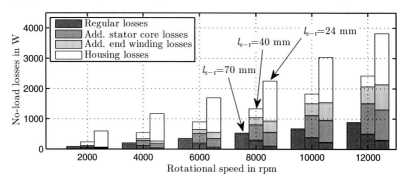

Figure 7.27: Summary of no-load losses

In the second place, the resultant efficiency map of the ADR-BLDC machine and also the efficiency map without the additional losses are shown in Fig. 7.28. When these efficiency maps are compared, it is obvious that the influence of the additional losses on the machine efficiency is considerable high. For example, at rotational speeds higher than $10000\,\text{rpm}$ these losses cause a $5 - 8\%$ decrease in the efficiency around $50\,\text{kW}$.

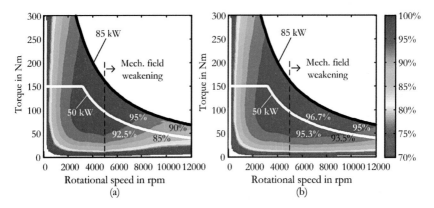

Figure 7.28: Efficiency map of the ADR-BLDC machine except the inverter and gearbox losses a) resultant and b) without the additional losses

The important findings presented in this chapter can be summarized as follows:

- The regular losses at a given rotational speed, including core and PM losses, approximately decrease linearly with $l_{\mathrm{s-r}}$ based on the made assumptions and definitions.

- The losses in the middle stator region due to the phase currents are small compared to the other losses.

- The additional end winding losses increase as $l_{\mathrm{s-r}}$ decreases but this increase is low once the rotor overhang is longer than the axial end winding length.

- The additional stator core losses at a given rotational speed stay almost constant for $l_{\mathrm{s-r}} < 50\,\mathrm{mm}$. Moreover, the eddy current losses due to the axial flux compose a big portion of these losses.

- The additional losses in the construction parts can be critical for ADR-BLDC machines. For example, the losses in one side of the housing are around 700 W at 9000 rpm. However, the additional losses in the construction parts are not directly related to the electromagnetic principle of ADR-BLDC machines. Therefore, they can be reduced by proper measures.

The analyses in this chapter show that additional losses can be critical depending on the machine design. Therefore, they must be considered in the design of PM electric machines designed for a mechanical field weakening operation. Due to the importance of this subject, design measures to decrease additional losses are discussed in the following chapter.

8 Recommendations

The analyses carried out on the prototype ADR-BLDC machine result in novel knowledge on the performance of ADR PM brushless machines. This chapter includes the important design rules and recommendations that are concluded from these results. In the first section, possible design measures to reduce additional losses in the mechanical field weakening range are examined. In the second section, speed limits of PM brushless machines with an axially displaceable rotor are analyzed. Finally, the last section summarizes important design aspects for ADR PM brushless machines.

8.1 Reduction of Additional Losses

The losses in the magnetic circuit of an ADR PM brushless machine do not decrease when reducing the active axial length of the machine; but additional losses arise. The results from the previous chapter show that these losses can be critically high. Therefore, possible design measures to reduce additional losses are discussed in this section.

8.1.1 Additional End Winding Losses

Additional end winding losses of the analyzed ADR-BLDC machine are high (up to 800 W) at high rotational speeds, despite the thin diameter (0.3 mm) of the implemented Litz wire. Some geometrical changes to reduce these losses have already been discussed in section 7.2.2. Additional design measures are discussed in the following text.

Additional end winding losses can be reduced by bending the end winding conductors away from the rotor overhang. This measure is easy to implement. Therefore, it should be applied, if possible. However, its applicability can be restricted due to the limited flexibility of the end winding conductors. For example, due to the very compact end winding design of the ADR-BLDC machine, the end winding conductors can only be bent to a small extent.

A similar effect as bending the end winding conductors can be achieved by implementing the conductors at some distance to the PMs. According to the magnetic flux density distribution results from Fig. 5.22, the magnetic flux density values close to the rotor overhang are much higher than the magnetic flux density values in the other end winding parts. Therefore, if the end winding conductors are placed out of this region with relatively high magnetic flux density, it is possible to reduce the end winding losses. In order to examine the distribution of the additional end winding losses, the end winding region of the ADR-BLDC machine is partitioned into 6 layers, as shown in Fig. 8.1a. Fig. 8.1b shows the losses in each of these layers as a percentage of the total additional end winding losses at the corresponding axial stator/rotor overlap length. As can be seen, more than 60% of the additional end winding losses are dissipated in the first layer. These results show that the additional end winding losses can considerably be decreased by slightly increasing the distance between the end winding conductors and the rotor overhang.

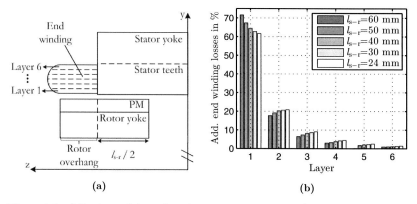

(a)   (b)

Figure 8.1: a) Partition of the end winding region into 6 layers, **b)** distribution of additional end winding losses among the layers as a percentage of the total additional end winding losses at the corresponding axial stator/rotor overlapping length l_{s-r}

Additionally, the end winding conductors can be shielded from the rotor overhang flux by implementing a stationary conductive plate directly under the end winding conductors. This method can be applied to reduce the flux penetrating into the end winding region. However, the losses in the conductive plate are expected to be critical at high rotational speeds. Therefore, this method is not appropriate to shield the PM leakage flux directly over the PMs. Instead of shielding the flux penetrating into the end winding region, a magnetic plate that rotates with the rotor can also be implemented to divert the rotor overhang flux from the end winding conductors. An important factor that limits the application of this measure is the narrow construction space between the rotor overhang surface and the end winding conductors.

The time-varying PM flux in the end winding region increases the phase flux linkage and therefore limits the achievable speed range in some degree with a limited axial rotor displacement range. Therefore, it is possible to extend the mechanical field weakening range by reducing the flux in the end winding regions.

8.1.2 Additional Stator Core Losses

As discussed in section 7.3.5, eddy current losses due to axial flux components in the stator core are mainly responsible for the additional stator core losses. They increase proportional to square of rotational speed; the paths of in-plane eddy currents are only restricted by the dimensions of the electric machine. Therefore, these losses can be decreased noticeably by reducing the fundamental frequency of the BLDC machine. Since the axial flux is much higher in the stator teeth, as shown in Fig. 7.11, the dimensions of the stator teeth have an important impact on the axial flux eddy current losses. Moreover, these losses can be reduced by reducing the axial flux in the stator and by increasing the resistance of the in-plane eddy current paths. These measures are examined in the following text.

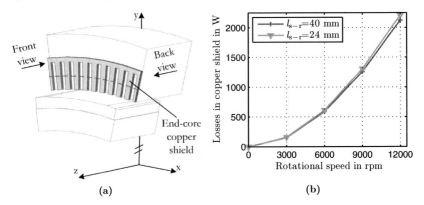

(a)   (b)

Figure 8.2: a) End-core copper shield, **b)** total losses in the end-core copper shields that are implemented at both sides of the stator core at l_{s-r} equal to 40 mm and 24 mm

In turbogenerators, end-core flux shields are used to screen the axial flux in the end regions of the stator core. Due to the good electrical conductivity of copper, copper shields are usually used, as discussed in [104] and [105]. The end-core copper shields can also be implemented in ADR PM brushless machines in order to reduce the additional stator core losses in the outer regions of the stator. Especially the stator teeth should be shielded. However, the stator core losses due to the inner fringing effects do not change. In order to test the applicability of this method, the losses in a stationary end-core copper shield are calculated by 3-D FEM transient analysis. Fig. 8.2a shows the construction of an end-core shield. As can be seen, it covers the stator teeth and only a small part of the stator yoke. Fig. 8.2b shows the total losses calculated in the end-core copper shields that are implemented at both sides of the stator core. Moreover, one of the end-core shields are divided into two parts from the radial middle of the teeth, in order to examine the loss distribution; it is found out that more than 80% of the losses are dissipated in the lower part because of the higher flux density values in this region. This can be seen in Fig. 8.3, which shows the magnetic flux density distribution in the end-core copper shield. These results show that the losses in the end-core copper shields are higher than the determined additional stator core losses that are shown in Fig. 7.24. Consequently, this method is not appropriate to reduce the additional losses of the ADR-BLDC machine. Therefore, the effect of the end-core copper shield on the magnetic flux distribition in the stator core is not analyzed in detail. On the other hand, this method can be interesting for some machine designs that have high axial flux losses in their stator cores, e.g. electric machines with relatively wide stator teeth.

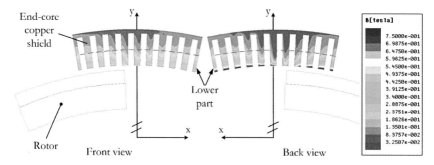

Figure 8.3: Magnetic flux density distribution on the surface of the end-core copper shield at 12000 rpm and l_{s-r}=24 mm

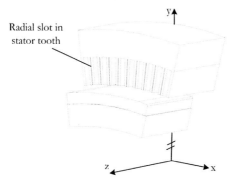

Figure 8.4: Radial slots (slits) in stator teeth

The resistance of the stator material in the x-y plane can be increased by cutting radial slots (slits), which is a design feature commonly employed in turbogenerators, in order to reduce the axial leakage flux losses in stators [104]. Fig. 8.4 shows the radial slots in the stator teeth of the ADR-BLDC machine. Cutting radial slots in the stator teeth is an effective way of decreasing the additional stator core losses by slightly affecting the performance of the electric machine. Therefore, it should be applied, if possible. However, the application of this method is limited, if the stator teeth are narrow like with the analyzed ADR-BLDC machine.

8.1.3 Additional Losses in Construction Parts

Finally, design measures to reduce additional losses in the construction parts are examined. These losses can be reduced by changes in the related construction parts without affecting the functionality of the electric machine. At the end of section 7.2.2, possible changes in the housing geometry to reduce these losses are discussed. These can be similarly applied to other construction parts that are exposed to a time varying magnetic field. However, this is a complex procedure, since first the affected construction parts must be identified and then these parts must be constructed considering their mechanical strength as well as the additional losses. Alternatively, the rotor overhang leakage flux can be directed away from the construction parts. This can be realized by changing the rotor and stator geometry as well as by implementing new parts. These possibilities are discussed in the following text.

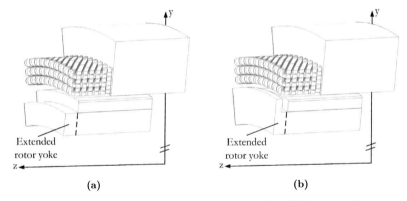

(a)   (b)

Figure 8.5: Extended rotor yoke designs applied to realize PM short-circuiting

In the first place, the rotor yoke of a SMPM rotor brushless machine can be designed in a way that the rotor overhang leakage flux is diverted away from the construction parts. Fig. 8.5 shows two possible rotor yoke designs applied to the ADR-BLDC machine, where the rotor yoke is axially extended. This change in the rotor construction is called the PM short-circuit, since the rotor overhang PM flux is short-circuited with the extended rotor part. Moreover, the end plates used to squeeze the rotor core laminated sheets, if the rotor is made of electric steel, can also be utilized to implement a PM short-circuit. A possible design of a flux diverting rotor end plate is shown in Fig. 8.6a. The extended rotor part and the rotor end plate must be designed with respect to their effect in the base speed range to avoid a reduction of the air-gap flux. This can be achieved by choosing a rotor yoke with an axial space between PM and extended rotor yoke, as shown Fig. 8.5b, and a rotor end plate consisting of a non-magnetic part and a magnetic part, as shown in Fig. 8.6a.

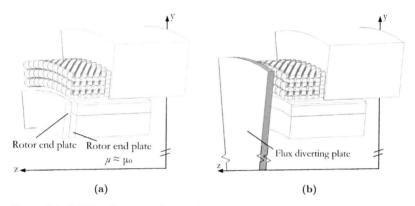

(a)   (b)

Figure 8.6: Additional parts implemented to divert the rotor overhang flux: a) rotor end plate consisting of a non-magnetic part and a magnetic part, b) flux diverting plate between housing and rotor

The design in Fig. 8.5b is examined in detail to find out the potential of a PM short-circuit approach in reducing additional losses in the housing of the ADR-BLDC machine. Fig. 8.7 shows the magnetic flux density distribution in the y-z plane. As can be seen, a significant part of the rotor overhang flux closes over the extended part of the rotor yoke. Additionally, the calculated losses *without* and *with* PM short-circuit are compared in Fig. 8.8. Accordingly, with this simple measure the losses in the housing are reduced to 29% of the calculated losses *without* PM short-circuit. These primary results show the high potential of implementing a PM short-circuit.

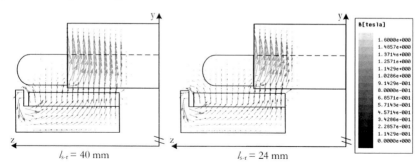

$l_{s\text{-}r} = 40$ mm $l_{s\text{-}r} = 24$ mm

Figure 8.7: Magnetic flux density vector in the y-z plane with PM short-circuit

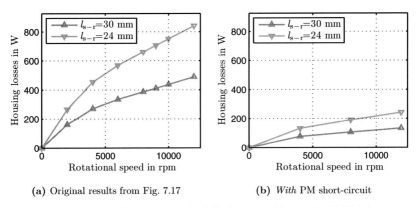

(a) Original results from Fig. 7.17   (b) *With* PM short-circuit

Figure 8.8: Losses simulated in one side of the housing *without* and *with* PM short-circuit

In case of an exterior rotor topology, as applied in [63], the PM short-circuit is automatically implemented. As can be seen in Fig. 8.9, the PM leakage flux can close over the high permeable shaft connection of the rotor yoke. Consequently, the other construction parts are shielded from the time-varying PM flux.

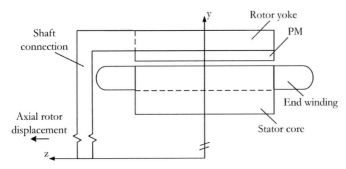

Figure 8.9: Outline of an exterior rotor BLDC machine

Similar to the PM short-circuit approach, a plate made of a material with a high magnetic permeability that rotates with the rotor shaft can be implemented between the PM rotor and the housing, as shown in Fig. 8.6b. This method is expected to be more effective than the PM short-circuit approach, since a flux diverting plate can be designed more freely. On the other hand, its complicated mechanical design, the required additional space and the additional weight are the drawbacks of this approach.

It is favorable to direct the rotor overhang leakage flux by those machine parts rotating with the rotor to avoid extra losses. Therefore, this kind of design measures is discussed up to this point. Additionally, the stator yoke can be extended over the end winding conductors, in order to divert the rotor overhang flux. Because of the relatively long axial distance between rotor surface and stator yoke, this measure is expected to have little influence on the magnetic flux in the side parts of the housing, where the losses are highest (Fig. 7.25). However, it can be applied, if the magnetic flux in the construction parts above the stator shall be diverted. Moreover, the stator yoke can be extended beyond the end winding conductors, which can make the production of the stator and the implementation of the winding more difficult. Similarly, additional stationary plates made of a material with high magnetic permeability can be attached at both sides of the electric machine. The benefits of these measures are limited due to the extra losses in the additional flux diverting stationary parts. Therefore, they are not examined in detail.

8.2 Limits

Mechanical field weakening is an alternative method for extending the speed range of PM brushless machines that have a limited electrical field weakening range. This applies to electric machines with high phase flux linkage and relatively low phase inductance. On the other hand, the mechanical field weakening range is also limited due to the limited DC-link voltage and phase current as well as due to the limited construction space, as discussed in section 2.2.3. In this section, the speed limits achievable with electrical, mechanical and combined field weakening methods are compared to analyze the application area of the mechanical field weakening method and the advantages of a combined field weakening method.

To examine the limits of the electrical and mechanical field weakening methods, the speed limits of five machine designs are analyzed with the help of analytical calculations neglecting the magnetic saturation. For these analyses, three-phase BLAC machines are chosen since their extended speed ranges can easily be approximated. All of these designs have the same phase flux linkage due to PM flux, phase-phase voltage, maximum phase current, pole pair number, peak power and maximum speed specification as the analyzed ADR-BLDC machine (table 8.1). The only difference between these designs is their phase inductances, which vary from $75\,\mu H$ to $350\,\mu H$. Only the phase inductances are varied since the other machine parameters are basically determined by the torque and supply specifications, and the inductances have an important influence on the field weakening operation. Moreover, the choice of a magnetically homogeneous rotor design ($L_d = L_q$) is necessary to be able to compare the field weakening methods since their speed ranges primarily depend on the different inductance components. At this point, it is important to note that each of these machine designs may have different design properties such as construction volume, rotor type, torque/power density. The analyses in this section do not aim at specifying an optimal machine design but at examining the operational limits of the field weakening methods.

$\hat{u}_\mathrm{s}$	$\hat{i}_\mathrm{p}$	n_p	$\hat{\Psi}_\mathrm{PM}$	P_peak	specified n_max
300 V	250 A	8	0.100 Vs	85 kW	12000 rpm

Table 8.1: Common properties of exemplary machine designs

Exemplary machine designs		Elec. FW	Mech. FW	
Design	$L_\mathrm{d}{=}L_\mathrm{q}$	$n_\mathrm{max\_at\_P_{peak},elec}$	$n_\mathrm{max\_at\_P_{peak},mech}$	min. $\hat{\Psi}_\mathrm{PM}$
1	75 μH	4100 rpm	12500 rpm	0.022 Vs
2	150 μH	5000 rpm	6200 rpm	0.043 Vs
3	225 μH	6600 rpm	4200 rpm	0.065 Vs
4	300 μH	10700 rpm	3100 rpm	0.086 Vs
5	350 μH	20200 rpm	2700 rpm	0.100 Vs

Table 8.2: Maximum rotational speeds at P_peak only achievable with electrical field weakening method and mechanical field weakening method without construction space limitations

The maximum rotational speeds only achievable at P_peak with electrical field weakening $n_\mathrm{max\_at\_P_{peak},elec}$ and alone mechanical field weakening $n_\mathrm{max\_at\_P_{peak},mech}$ without construction space limitations are given in table 8.2. These values are calculated neglecting the voltage drop on the phase resistance and assuming that the phase inductances are constant in the whole operating range by using the equations given in sections 2.2.2 and 2.2.3. In case there are no construction space limitations, the phase flux linkage can be arbitrarily reduced by adjusting the axial rotor position. Accordingly, in contrary to the electrical field weakening, $n_\mathrm{max\_at\_P_{peak},mech}$ does not depend on the phase flux linkage value at the rotor base position. Therefore, the amplitudes of Ψ_PM at $n_\mathrm{max\_at\_P_{peak},mech}$ are provided in the last column of table 8.2. These results show that the range of electrical field weakening extends, and mechanical field weakening decreases as the phase inductance is increased. The electrical field weakening range extends, since the ratio between the phase flux linkage and the phase inductance reduces. On the other hand, the mechanical field weakening range decreases, since the voltage drop on the phase inductance at a given operating point increases. Due to the high flux linkage, the electrical field weakening range is limited, and the specified maximum rotational speed 12000 rpm cannot be reached for a phase inductance lower than 300 μH. On the other hand, the maximum rotational speed with mechanical field weakening is also limited; only the design with 75 μH can reach the specified maximum rotational speed. The speed ranges of the designs can be extended beyond $n_\mathrm{max\_at\_P_{peak},mech}$, only if the power is reduced, as discussed in section 2.2.3.

Additionally, the construction space can be the factor that limits the achievable speed range of the mechanical field weakening method. This applies if the minimum flux linkage limit cannot be reached in the axial rotor displacement range. To illustrate the speed limits due to the construction space limitation, design 1 (75 μH) is chosen due to its wider useful axial displacement range. Fig. 8.10 shows the speed limits of this design for varying minimum phase flux linkage values. As expected, the achievable maximum speed highly depends on the minimum value of the phase flux linkage.

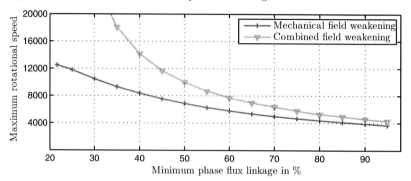

Figure 8.10: Speed limits of design 1 with mechanical and combined field weakening methods with limited axial rotor displacement range

As a result, it can be concluded that electrical or mechanical field weakening methods may not be sufficient to reach a specified maximum rotational speed alone. Alternatively, these field weakening methods can be combined, as suggested in section 6.2.2. The speed limits of design 1 with the combined field weakening method are also shown in Fig. 8.10. As can be seen, the combined field weakening method results in significantly higher maximum rotational speeds than the electrical and mechanical field weakening methods, especially at low $\hat{\Psi}_{\mathrm{PM}}$ values. This is because of the collaborative operation between the field weakening methods. For example, if the phase flux linkage is reduced with mechanical field weakening, electrical field weakening becomes more effective, since the ratio between phase flux linkage and phase inductance increases. Therefore, the combined field weakening strategy offers an optimal solution.

The speed limits of the exemplary machine designs with combined field weakening are given in table 8.3. All exemplary designs fulfill the maximum speed specification. Moreover, the rotational speeds achievable with combined field weakening without construction space limitations are higher than the specified maximum speed. This shows that phase flux linkages are not required to be reduced to their lowest possible values. Therefore, the required minimum limits of the phase flux linkages are determined so that the maximum achievable rotational speed is equal to the rotational speed specification, which is 12000 rpm. These flux linkage values are listed additionally in the third column of table 8.3. Consequently, it is possible to reduce the required axial rotor displacement range, if the combined field weakening method is applied.

Moreover, a combined field weakening strategy introduces another degree of freedom to the control which can be used to enhance the machine performance. For instance, the optimum axial rotor position can be determined considering the losses. Accordingly, first

Design	Combined electrical and mechanical field weakening			
	min. Ψ_{PM}	$n_{max\_at\_P_{peak},com}$	min. Ψ_{PM}	$n_{max\_at\_P_{peak},com}$
1	0.022 Vs	88200 rpm	0.045 Vs	12000 rpm
2	0.043 Vs	44100 rpm	0.063 Vs	12000 rpm
3	0.065 Vs	29400 rpm	0.079 Vs	12000 rpm
4	0.086 Vs	22000 rpm	0.097 Vs	12000 rpm
5	0.100 Vs	20200 rpm	-	-

Table 8.3: Maximum rotational speeds at P_{peak} achievable with combined field weakening method

electrical field weakening is applied in case of high additional losses in the mechanical field weakening range. Furthermore, the proposed combined field weakening method in section 6.2.2 that is applied to enhance the phase current characteristics is another application example.

The analyses in this section are carried out assuming that the phase inductances are constant, and only the amplitude of the phase flux linkage reduces as the active length of the machine is reduced. However, once the rotor is axially displaced, the self-inductances are expected to slightly increase, whereas the mutual inductances slightly decrease and the flux linkage (back EMF) waveform changes due to additional flux in the end winding regions. These changes must also be considered in order to calculate the speed limits more accurately.

Speed limits of the exemplary BLAC machine designs with sinusoidal back EMF and phase current waveforms have been examined so far. The simulation results from section 6.2.2 indicate that a combined field weakening method is similarly very advantageous for BLDC machines. In this case, the phase advance field weakening method has to be applied due to the non-sinusoidal back EMF and phase current characteristics.

In addition to the speed limits, the short-circuit current in case of inverter and winding faults is another design criterion for traction applications. The short-circuit current at a terminal short-circuit condition is limited by the phase reactance. Therefore, the short-circuit current can be high, if the phase flux linkage is high and the phase inductance relatively low. However, the short-circuit current can be reduced effectively by adjusting the active machine length. Therefore, the required axial rotor displacement range must be determined considering the short-circuit current in addition to the rotational speed range.

To sum up, the limits of the electrical and mechanical weakening methods highly depend on the machine properties and they may not be sufficient to realize the specified speed range. However, when these field weakening methods are combined, the speed range of an electric machine extends beyond the limits of each of these field weakening methods. The analyses in this section show that combined electrical and mechanical field weakening is an effective method for extending the speed range of PM brushless machines. Therefore, it should be applied to realize a wide speed range with a reduced axial rotor displacement range.

8.3 Design Considerations

One of the biggest drawbacks of mechanical field weakening in traction applications is the required construction space for the axial rotor displacement. Therefore, the axial rotor displacement must be limited by the available construction space. If the construction space is the limiting factor, electric machines with a short axial length are more favorable for mechanical field weakening. Furthermore, the construction space available under the end winding conductors can be utilized for the axial rotor displacement in an interior rotor machine. This suggests the usage of a rotor with two symmetrical parts, which are axially displaced apart from each other, in the mechanical field weakening range in order to utilize the construction space at both sides.

The effectiveness of mechanical field weakening gains importance, in case the axial rotor displacement range is limited. The results of chapter 5 and 6 show that the additional flux components in the end winding regions and in the stator core reduce the speed range achievable with a limited construction space. Because of these flux components, the amplitude of flux linkage does not decrease proportional to the axial stator/rotor overlap length. For example, the magnitude of the flux linkage of the analyzed ADR-BLDC machine is halved by reducing the axial stator/rotor overlap length to almost one third. The influence of the additional flux becomes critical, as the axial length of the machine decreases. On the other hand, the required axial displacement to overcome these effects is small, too.

The implementation of an undivided rotor has some important advantages compared to a rotor geometry with two symmetric parts. If the rotor consists of two parts, undesired effects such as the increase in the flux linkage due to additional flux components, and additional losses in the end winding conductors and in the stator core are at least twice as high as in case of an undivided rotor. Moreover, additional losses in the construction parts only occur in one side of a machine with an undivided rotor, which makes it easier to reduce the losses in the construction parts. Therefore, if the construction space at one side of the electric machine is sufficient for axial rotor displacement, an undivided rotor geometry is favorable. Otherwise, there is a trade-off between the required additional construction space and the reduction of undesired effects. Additionally, possible differences in dynamics of the displacement mechanism between undivided and divided rotor topologies must be considered.

Three additional loss mechanisms, i.e. additional end winding losses, additional stator core losses and additional losses in construction parts, are defined in the mechanical field weakening range. An important outcome from loss analysis in chapter 7 is that these additional losses can be considerably high, unless measures to reduce them are implemented. Due to the importance of this subject, possible design measures to decrease additional losses are examined in section 8.1.3. These analyses show that additional losses can significantly be decreased with an appropriate design. Some of these measures are easy to realize, as in the case of housing losses. Additional losses must certainly be examined in the early design phase to be able to take appropriate measures.

The results for the additional end winding losses from Fig. 7.23 and the additional stator core losses from Fig. 7.24 show that these losses increase almost proportional to square of rotational speed. Consequently, reducing the number of pole pairs n_p can result in a significant reduction in additional losses. For example, if n_p of the ADR-BLDC machine is reduced from 8 to 5, the additional end winding losses and the additional stator core losses are expected to decrease almost by 60%. Based on this, n_p must carefully be determined considering its effect on the required current ratings of the electrical drive and the machine design, such as the end winding design and the thickness of the stator yoke, as well as on the additional losses.

For an effective and efficient field weakening operation, a combination of electrical and mechanical field weakening methods is advantageous, as discussed in sections 6.2.2 and 8.2. With this method, the speed range of mechanical field weakening can be extended, and the construction space required for axial rotor displacement can be reduced.

9 Conclusion and Future Work

PM brushless machines are used in vehicle traction applications because of their high torque/power density and their high efficiency at full load. However, the constant PM excitation limits their high speed performance. Therefore, methods to extend the rotational speed range of PM brushless machines without decreasing the torque/power density and efficiency are presented in various researches. These studies deal either with PM brushless machine topologies that have a better high speed performance when standard field weakening methods are applied or with novel PM brushless machine topologies that are developed for non-conventional field weakening approaches. One of these non-conventional field weakening approaches is the mechanical field weakening method, in which the axial active length of the PM brushless machine is mechanically decreased by axially displacing the rotor relative to the stator. This method can be applied to conventional electric machines, if an additional rotor displacement mechanism is implemented, and it is an effective method to extend the speed range of PM brushless machines with a limited electrical field weakening range. Therefore, the performance of this mechanical field weakening method is analyzed in this study. In these analyses, a prototype multi-phase ADR-BLDC machine is used as test object.

In this study, the mechanical field weakening method is examined under three main aspects. These are the dependency of the electrical machine parameters on the axial rotor displacement, the effects of the axial rotor displacement on the machine's operational characteristics and the losses arising due to stator/rotor misalignment.

To begin with, the electrical machine parameters such as flux linkage and induced back EMF waveforms of the ADR-BLDC machine are calculated by 3-D FEM analysis. The results show that the effectiveness of the mechanical field weakening is limited due to additional flux components in the stator core and in the end winding regions. The additional flux components arise due to stator/rotor misalignment and affect the torque coefficient of the ADR-BLDC machine, as well. The analysis shows that it is possible to determine the torque constant in the field weakening range by using the calculated back EMF characteristics in case of a negligible magnetic saturation. Furthermore, a model reduction approach used to determine the self- and mutual inductance values is proposed due to the high computational capacity required for transient 3-D FEM calculations. In this method, the electric machine is divided into four axial sections according to magnetostatic field analysis. The three-dimensional magnetic field distribution in these sections is approximated by three 2-D FEM models and a simplified 3-D FEM model. The inductance values calculated with this approach are validated by experimental results. As a result, it is found out that the self-inductance values slightly increase with the axial rotor displacement, whereas the mutual inductance values slightly decrease.

In order to analyze the operational characteristics, the electric machine parameters are implemented in a dynamic simulation model in a second step. The model enables very fast dynamic simulations considering the non-linear behavior of the magnetic materials as well as the effects of the axial rotor position. The validity of the dynamic simulation model is verified for two extreme axial rotor positions by measurement results. By using this model, operational characteristics of the ADR-BLDC machine with pulse amplitude modulation control are simulated. These analyses show that the operational characteristics of the ADR-BLDC machine mainly change in the field weakening range due to the change in the back EMF waveform and the decrease in mutual inductance values. Moreover, speed limits of the ADR-BLDC machine are compared for two field weakening strategies. As a result, a combined electrical and mechanical field weakening method is proposed. In this case, electrical field weakening is used to enhance the speed range as well as the phase current characteristics of the ADR-BLDC machine.

In a final step, the losses in the mechanical field weakening range are analyzed by using numerical and experimental methods. First, the losses in the base speed range are determined, and their dependency on the axial rotor position is explained with the help of 3-D magnetic field distribution results. As a result, the regular core losses and the PM losses decrease as the axial active length of the machine is reduced, but the copper losses at a given output torque increase, since the machine torque constant decreases in the mechanical field weakening range. On the other hand, three additional loss mechanisms arising due to stator/rotor misalignment are identified. These are additional end winding losses, additional stator core losses and additional losses in the construction parts. Numerical methods are developed to calculate these additional losses. The analyses show that only the additional stator core losses slightly increase with load, whereas the other additional losses can be assumed to be load-independent. The calculation results are validated and extended by the no-load loss measurement results. These experiments are carried out with different stators and rotors in order to be able to separate the losses. An important outcome of loss analyses is that the performance of the mechanical field weakening method is limited by the losses at high rotational speeds. For example, at 9000 rpm and maximum rotor displacement, the additional end winding losses are around 800 W, the losses in one side of the housing are around 700 W, and the additional stator core losses are around 600 W. The total of these additional losses results in a 4.2% decrease in the efficiency of the ADR-BLDC machine at 50 kW. This result indicates the need for a design optimization to reduce additional losses.

The analyses in this study deliver important findings on the performance and limits of the mechanical field weakening method. In the light of these results, recommendations for ADR PM brushless machines are provided. The measures to reduce additional losses constitute a remarkable part of these recommendations. In addition to conventional measures, new design features are proposed and analyzed. As a result, it is found out that additional losses can be reduced significantly. Moreover, the field of application of mechanical field weakening is examined by comparing the speed ranges achievable with the electrical, mechanical and combined field weakening methods. These analyses reveal that mechanical field weakening is favorable, if the speed range with electrical field weakening is limited. On the other hand, a combined field weakening method has to be implemented to realize a wide speed range with a limited axial rotor displacement range.

The functionality of the analyzed mechanical field weakening method is proven, and aspects that have to be considered in the design process of such PM brushless machines are presented. The additional construction space needed for axial rotor displacement can be reduced, if the axial length of the electric machine is kept short, the construction space under the end winding regions is utilized for axial rotor displacement and a combined electrical and mechanical field weakening method is applied. The practical application of this field weakening method is questionable not because of the required additional construction space, but because of the high additional losses. Therefore, in the design process of PM brushless machines with axially displaceable rotor, these losses must be considered, and measures to reduce them must be taken.

The findings of this study can be used in future studies to enhance the performance of PM brushless machines. For example, the design changes proposed to reduce the additional losses can be optimized and tested experimentally and the application of the combined field weakening method can be optimized by considering the system efficiency. Moreover, the effect of the rotor topology on the performance of the mechanical field weakening method is also an important design question, since rotor topologies have different magnetic field distributions in the rotor overhang region. Therefore, additional flux components arising due to stator/rotor misalignment must be evaluated with different rotor topologies to find out the most appropriate rotor design. Based on the fact that a significant part of the stator of an ADR PM brushless machine is exposed to a 3-D flux, the application of soft magnetic composite materials would be a future research subject. Additionally, the mechanical design of the axial rotor displacement mechanism has to be investigated in future studies. A robust and compact mechanical field weakening mechanism has to be developed considering radial and axial magnetic forces acting between stator and rotor.

A Measuring Instruments

The measuring instruments used in this study are listed in table A.1. Some of these devices are directly used to measure the machine parameters; the others are implemented in the measurement setups.

Quantity	Used measuring instrument	Drag T.	Load T.
Angular position	Digital Hall effect switch TLE4946H [106]		X
Current	Current transducer Ultrastab 867-700I* [107]		X
	Multi channel transducer system MCTS* [108]		X
DC resistance	Micro ohmmeter MR5-200C [109]		
Impedance	LCR bridge HM8118 [110]		
Rotational speed	Hall effect gear teeth sensor GT101DC** [111]	X	
	Pulse generator***		X
Temperature	KTY 83/110 [112]	X	X
Torque	Torque sensor T12*** [113] Nominal torque = 100 Nm	X	
	Nominal torque = 2500 Nm		X
Voltage	Power analyzer WT3000 [114]	X	X

Table A.1: Measuring instruments
* Recorded via power analyzer WT3000 with 200 kHz sampling rate
** Recorded via speed input of the power analyzer WT3000
*** Recorded via test bench

There are two measurement setups:

- **Drag test setup:** This setup is used to perform open circuit drag tests. At first, the back EMF waveforms are measured with this setup. The results of these tests are used in section 5.4 to validate the back EMF characteristics that are simulated with the FEM analysis. Moreover, this setup is used to measure the no-load losses in the drive system as explained in section 7.3. Detailed information about this test setup and the devices under test are provided in the corresponding sections.

- **Load test setup:** Load tests are carried out to examine the operational characteristics of the ADR-BLDC machine drive system both in the base speed and mechanical field weakening range. This test setup is introduced in section 7.3.

149

Bibliography

[1] J. Larminie and J. Lowry, *Electric Vehicle Technology Explained*. Wiley, 2012.

[2] S. Chu, *Critical Materials Strategy*. DIANE Publishing Company, 2011.

[3] L. Zepp and J. Medlin, "Brushless permanent magnet motor or alternator with variable axial rotor/stator alignment to increase speed capability," Apr. 29 2003, US Patent 6,555,941.

[4] J. Medlin and L. Zepp, "Brushless permanent magnet motor or alternator with variable axial rotor/stator alignment to increase speed capability," Sep. 18 2003, WO Patent App. PCT/US2003/006,897.

[5] H. Wöhl-Bruhn, "Synchronmaschine mit eingebetteten Magneten und neuartiger variabler Erregung für Hybridantriebe," Ph.D. dissertation, Technische Universität Braunschweig, 2010.

[6] H. Amecke, C. Besch, E. Bostanci, S. Martin, C. Mertens, Z. Neuschl, R. Plikat, O. Rauch, E. Schulze, H. Strauss, and H. Wetzel, "Leistungsdichte E-Maschine," Volkswagen AG, Tech. Rep., 2011. [Online]. Available: http://www.erneuerbar-mobil.de/projekte/ foerderprojekte-aus-dem-konjunkturpaket-ii-2009-2011/pkw-feldversuche/ abschlussberichte/abschlussbericht-lde-m.pdf

[7] Z. Zhu and D. Howe, "Electrical machines and drives for electric, hybrid, and fuel cell vehicles," *Proceedings of the IEEE*, vol. 95, no. 4, pp. 746 –765, 2007.

[8] C. Chan, "The state of the art of electric and hybrid vehicles," *Proceedings of the IEEE*, vol. 90, no. 2, pp. 247–275, 2002.

[9] K. Chau, C. Chan, and C. Liu, "Overview of permanent-magnet brushless drives for electric and hybrid electric vehicles," *Industrial Electronics, IEEE Transactions on*, vol. 55, no. 6, pp. 2246–2257, 2008.

[10] A. Sorniotti, M. Boscolo, A. Turner, and C. Cavallino, "Optimization of a multi-speed electric axle as a function of the electric motor properties," in *Vehicle Power and Propulsion Conference (VPPC), 2010 IEEE*, 2010, pp. 1–6.

[11] Z. Rahman, E. Ehsani, and K. Butler, "An investigation of electric motor drive characteristics for EV and HEV propulsion systems," Texas A&M University, Tech. Rep., 2000. [Online]. Available: http://psalserver.tamu.edu/main/papers/009% 20Rahman%20Ehsani%20Butler.pdf

[12] T. Finken, M. Felden, and K. Hameyer, "Comparison and design of different electrical machine types regarding their applicability in hybrid electrical vehicles," in *Electrical Machines, 2008. ICEM 2008. 18th International Conference on*, 2008, pp. 1–5.

[13] M. Zeraouila, M. E. H. Benbouzid, and D. Diallo, "Electric motor drive selection issues for HEV propulsion systems: a comparative study," in *Vehicle Power and Propulsion, 2005 IEEE Conference*, 2005.

[14] M. Kamiya, "Development of traction drive motors for the toyota hybrid system," *IEEJ Transactions on Industry Applications*, vol. 126, pp. 473–479, 2006.

[15] D. O'Connell, "Tesla motors," Tesla Motors, 2010. [Online]. Available: http://www.electricdrive.org/index.php?ht=a/GetDocumentAction/i/15351

[16] Renault, "Electric motor 3CG," 2013. [Online]. Available: http://www.renault.com/en/innovation/gamme-mecanique/pages/moteur-electrique-3cg.aspx

[17] A. Vignaud and H. Fennel, "Efficient electric powertrain with externally excited synchronous machine without rare earth magnets using the example of the renault system solution," Continental, 2012. [Online]. Available: http://www.conti-online.com/www/download/automotive_de_de/general/powertrain/download/2012_wien_vortrag_uv.pdf

[18] Renault, "The 2010 paris motor show," 2010. [Online]. Available: http://www.renault.com/SiteCollectionDocuments/Communiqu%C3%A9%20de%20presse/en-EN/Pieces%20jointes/23471_RENAULT_-__GB_790FD118.pdf

[19] F. B. Group, "Bluecar, from drawing-board to reality," Geneva Motor Show Press Release, 2006. [Online]. Available: http://www.batscap.com/en/actualites/communiques/DossierDePresse-03-2006.pdf

[20] A. Parviainen, "Design of axial-flux permanent-magnet low-speed machines and performance comparison between radial-flux and axial-flux machines," Ph.D. dissertation, Lappeenranta University of Technology, 2005.

[21] S. Lee, "Development and analysis of interior permanent magnet synchronous motor with field excitation structure," Ph.D. dissertation, University of Tennessee - Knoxville, 2009.

[22] A. El-Refaie, "Fractional-slot concentrated-windings synchronous permanent magnet machines: Opportunities and challenges," *Industrial Electronics, IEEE Transactions on*, vol. 57, no. 1, pp. 107 –121, 2010.

[23] R. Krishnan, *Permanent Magnet Synchronous and Brushless DC Motor Drives*. Taylor and Francis Group, LCC, 2010.

[24] L. Parsa, "Performance improvement of permanent magnet AC motors," Ph.D. dissertation, Texas A&M University, 2005.

[25] A. El-Refaie and T. Jahns, "Impact of winding layer number and magnet type on synchronous surface PM machines designed for wide constant-power speed range operation," *Energy Conversion, IEEE Transactions on*, vol. 23, no. 1, pp. 53–60, 2008.

[26] S. Chaithongsuk, B. Nahid-Mobarakeh, J. Caron, N. Takorabet, and F. Meibody-Tabar, "Optimal design of permanent magnet motors to improve field-weakening performances in variable speed drives," *Industrial Electronics, IEEE Transactions on*, vol. 59, no. 6, pp. 2484–2494, 2012.

[27] T. Lipo and M. Aydin, "Field weakening of permanent magnet machines - design approaches," Electrical and Computer Engineering Department University of Wisconsin , Madison, Tech. Rep., 2004. [Online]. Available: http://lipo.ece.wisc.edu/2004pubs/2004_13.pdf

[28] H. Woehl-Bruhn, W.-R. Canders, and N. Domann, "Classification of field-weakening solutions and novel PM machine with adjustable excitation," in *Electrical Machines (ICEM), 2010 XIX International Conference on*, 2010, pp. 1–6.

[29] R. Owen, Z. Zhu, J. Wang, D. A. Stone, and I. Urquhart, "Review of variable-flux permanent magnet machines," *Journal of International Conference on Electrical Machines and Systems*, vol. 1, pp. 23–31, 2012.

[30] T. Schoenen, M. Kunter, M. Hennen, and R. De Doncker, "Advantages of a variable DC-link voltage by using a DC-DC converter in hybrid-electric vehicles," in *Vehicle Power and Propulsion Conference (VPPC), 2010 IEEE*, 2010, pp. 1–5.

[31] J. Estima and A. Marques Cardoso, "Efficiency analysis of drive train topologies applied to electric/hybrid vehicles," *Vehicular Technology, IEEE Transactions on*, vol. 61, no. 3, pp. 1021–1031, 2012.

[32] V. Ostovic, "Memory motors," *Industry Applications Magazine, IEEE*, vol. 9, no. 1, pp. 52–61, 2003.

[33] C. Yu and K. Chau, "Design, analysis, and control of DC-excited memory motors," *Energy Conversion, IEEE Transactions on*, vol. 26, no. 2, pp. 479–489, 2011.

[34] F. Li, K. Chau, C. Liu, and Z. Zhang, "Design principles of permanent magnet dual-memory machines," *Magnetics, IEEE Transactions on*, vol. 48, no. 11, pp. 3234–3237, 2012.

[35] C.-H. Zhao and Y.-G. Yan, "A review of development of hybrid excitation synchronous machine," in *Industrial Electronics, 2005. ISIE 2005. Proceedings of the IEEE International Symposium on*, vol. 2, 2005, pp. 857–862.

[36] Y. Amara, L. Vido, M. Gabsi, E. Hoang, A. Hamid Ben Ahmed, and M. Lecrivain, "Hybrid excitation synchronous machines: Energy-efficient solution for vehicles propulsion," *Vehicular Technology, IEEE Transactions on*, vol. 58, no. 5, pp. 2137–2149, 2009.

[37] X. Luo and T. A. Lipo, "A synchronous/permanent magnet hybrid AC machine," *Energy Conversion, IEEE Transactions on*, vol. 15, no. 2, pp. 203–210, 2000.

[38] T. Finken and K. Hameyer, "Study of hybrid excited synchronous alternators for automotive applications using coupled FE and circuit simulations," *Magnetics, IEEE Transactions on*, vol. 44, no. 6, pp. 1598–1601, 2008.

[39] D. Fodorean, A. Djerdir, l.-A. Viorel, and A. Miraoui, "A double excited synchronous machine for direct drive application - design and prototype tests," *Energy Conversion, IEEE Transactions on*, vol. 22, no. 3, pp. 656–665, 2007.

[40] A. Shakal, Y. Liao, and T. Lipo, "A permanent magnet AC machine structure with true field weakening capability," in *Industrial Electronics, 1993. Conference Proceedings, ISIE'93 - Budapest., IEEE International Symposium on*, 1993, pp. 19–24.

[41] F. Leonardi, P. McCleer, and T. Lipo, "The DSPM: An AC permanent magnet traction motor with true field weakening," Wisconsin Electric Machines and Power Electronics Consortium, Tech. Rep., 1997. [Online]. Available: http://lipo.ece.wisc.edu/1997pub/97-27.PDF

[42] K. Chau, J. Jiang, and Y. Wang, "A novel stator doubly fed doubly salient permanent magnet brushless machine," *Magnetics, IEEE Transactions on*, vol. 39, no. 5, pp. 3001–3003, 2003.

[43] K. Chau, Y. Li, J. Jiang, and C. Liu, "Design and analysis of a stator-doubly-fed doubly-salient permanent-magnet machine for automotive engines," *Magnetics, IEEE Transactions on*, vol. 42, no. 10, pp. 3470–3472, 2006.

[44] X. Zhu, M. Cheng, W. Zhao, C. Liu, and K. Chau, "A transient cosimulation approach to performance analysis of hybrid excited doubly salient machine considering indirect field-circuit coupling," *Magnetics, IEEE Transactions on*, vol. 43, no. 6, pp. 2558–2560, 2007.

[45] Z. Q. Zhu, Y. Pang, D. Howe, S. Iwasaki, R. Deodhar, and A. Pride, "Analysis of electromagnetic performance of flux-switching permanent-magnet machines by nonlinear adaptive lumped parameter magnetic circuit model," *Magnetics, IEEE Transactions on*, vol. 41, no. 11, pp. 4277–4287, 2005.

[46] J. T. Chen, Z. Q. Zhu, S. Iwasaki, and R. P. Deodhar, "A novel hybrid-excited switched-flux brushless AC machine for EV/HEV applications," *Vehicular Technology, IEEE Transactions on*, vol. 60, no. 4, pp. 1365–1373, 2011.

[47] W. Hua, Z. Zhu, M. Cheng, Y. Pang, and D. Howe, "Comparison of flux-switching and doubly-salient permanent magnet brushless machines," in *Electrical Machines and Systems, 2005. ICEMS 2005. Proceedings of the Eighth International Conference on*, vol. 1, 2005, pp. 165–170.

[48] J. A. Tapia, F. Leonardi, and T. A. Lipo, "Consequent pole PM machine with field weakening capability," in *IEEE International Conference on Electrical Machines and Drive*, 2001, pp. 126–131.

[49] J. Tapia, F. Leonardi, and T. Lipo, "Consequent-pole permanent-magnet machine with extended field-weakening capability," *Industry Applications, IEEE Transactions on*, vol. 39, no. 6, pp. 1704–1709, 2003.

[50] L. Ma, M. Sanada, S. Morimoto, Y. Takeda, and N. Matsui, "High efficiency adjustable speed control of IPMSM with variable permanent magnet flux linkage," in *Industry Applications Conference, 1999. Thirty-Fourth IAS Annual Meeting. Conference Record of the 1999 IEEE*, vol. 2, 1999, pp. 881–887.

[51] L. Ma, M. Sanada, S. Morimoto, and Y. Takeda, "Advantages of IPMSM with adjustable PM armature flux linkage in efficiency improvement and operating range extension," in *Power Conversion Conference, 2002. PCC-Osaka 2002. Proceedings of the*, vol. 1, 2002, pp. 136–141.

[52] R. Owen, Z. Zhu, J. Wang, D. Stone, and I. Urquhart, "Mechanically adjusted variable-flux concept for switched-flux permanent-magnet machines," in *Electrical Machines and Systems (ICEMS), 2011 International Conference on*, 2011, pp. 1–6.

[53] Z. Q. Zhu, M. M. J. Al-Ani, X. Liu, M. Hasegawa, A. Pride, and R. Deodhar, "Comparison of flux weakening capability in alternative switched flux permanent magnet machines by mechanical adjusters," in *Electrical Machines (ICEM), 2012 XXth International Conference on*, 2012, pp. 2889–2895.

[54] K. Baoquan, L. Chunyan, and C. Shukang, "A new flux weakening method of permanent magnet synchronous machine," in *Electrical Machines and Systems, 2005. ICEMS 2005. Proceedings of the Eighth International Conference on*, vol. 1, 2005, pp. 500–503.

[55] E. Nipp, "Alternative to field-weakening of surface-mounted permanent-magnet motors for variable-speed drives," in *Industry Applications Conference, 1995. Thirtieth IAS Annual Meeting, IAS '95., Conference Record of the 1995 IEEE*, vol. 1, 1995, pp. 191 –198.

[56] P. Otaduy, J. Hsu, and D. Adams, "Study of the advantages of internal permanent magnet drive motor with selectable windings for hybrid electric vehicles," Oak Ridge National Laboratory, Tech. Rep., 2007. [Online]. Available: http://www.osti.gov/scitech/servlets/purl/921779

[57] M. Swamy, T. Kume, A. Maemura, and S. Morimoto, "Extended high-speed operation via electronic winding-change method for AC motors," *Industry Applications, IEEE Transactions on*, vol. 42, no. 3, pp. 742–752, 2006.

[58] S. Sadeghi, L. Guo, H. Toliyat, and L. Parsa, "Wide operational speed range of five-phase permanent magnet machines by using different stator winding configurations," *Industrial Electronics, IEEE Transactions on*, vol. 59, no. 6, pp. 2621–2631, 2012.

[59] B. Kraßer, "Optimierte Auslegung einer Modularen Dauermagnetmaschine für ein Autarkes Hybridfahrzeug," Ph.D. dissertation, Technische Universität München, 2000.

[60] G. Zhou, T. Miyazaki, S. Kawamata, D. Kaneko, and N. Hino, "Development of variable magnetic flux motor suitable for electric vehicle," in *Power Electronics Conference (IPEC), 2010 International*, 2010, pp. 2171–2174.

[61] F. Liang, Y. Liao, and T. Lipo, "Field weakening for a doubly salient motor with stator permanent magnets," Sep. 1 1994, CA Patent App. CA 2,154,491.

[62] R. Himmelmann, M. Shahamat, and D. Halsey, "Permanent magnet dynamoelectric machine with axially displaceable permanent magnet rotor assembly," Jun. 10 2008, US Patent 7,385,332.

[63] W. Steiger, T. Böhm, and B.-G. Schulze, "Directhybrid - a combination of combustion engine and electric transmission," in *15. Aachener Kolloquium Fahrzeug und Motorentechnik*, 2006.

[64] B. Schulze and A. Dittner, "Generator-Motor-Kombination," Jul. 6 1995, DE Patent 4,408,719.

[65] Y. Kats, "Adjustable-speed drives with multiphase motors," in *Electric Machines and Drives Conference Record, 1997. IEEE International*, 1997, pp. TC2/4.1–TC2/4.3.

[66] J. Huang, M. Kang, J. Yang, H. Jiang, and D. Liu, "Multiphase machine theory and its applications," in *Electrical Machines and Systems, 2008. ICEMS 2008. International Conference on*, 2008, pp. 1–7.

[67] G. Müller, K. Vogt, and B. Ponick, *Berechnung elektrischer Maschinen*, ser. Elektrische Maschine Series.   Wiley VCH Verlag GmbH, 2007.

[68] M. Maerz, B. Eckardt, and A. Schletz, *Mechatronic integration of power electronics in hybrid powertrain components*.   Renningen-Malmsheim: Expert Verlag, 2007 (Haus der Technik Fachbuchreihe 80), 2007.

[69] F. Magnussen, P. Thelin, and C. Sadarangani, "Performance evaluation of permanent magnet synchronous machines with concentrated and distributed windings including the effect of field-weakening," in *Power Electronics, Machines and Drives, 2004. (PEMD 2004). Second International Conference on (Conf. Publ. No. 498)*, vol. 2, 2004, pp. 679 – 685.

[70] B. Schulze and G. Stöhr, "Wicklung für elektrische Maschinen," Jun. 1 2006, DE Patent App. DE200,410,055,608.

[71] J. S. Corporation, "Typical magnetic property curves for each grade of JFE super cores," 2013. [Online]. Available: http://www.jfe-steel.co.jp/en/products/electrical/catalog/f2e-001.pdf

[72] Z. Neuschl, "Computer aided experimental methods for determination of load independent iron losses in permanent magnet electrical machines with additionaly axial flux," Ph.D. dissertation, Brandenburg University of Technology, Cottbus, 2007.

[73] E. Bostanci, Z. Neuschl, R. Plikat, and B. Ponick, "No-load performance analysis of brushless dc machines with axially displaceable rotor," *Industrial Electronics, IEEE Transactions on*, vol. 61, no. 4, pp. 1692–1699, 2014.

[74] Raffmetal, *Aluminium alloy AlSi10Mg*, 2013. [Online]. Available: http://raffmetal. it/pdf/schede-leghe/en/EN-43400.pdf

[75] H. Vetter, "Mission- Profile bezogene PCC- Designs zur Integration in HEV- Converter," *Automobil-Elektronik*, 2007.

[76] A. Thomas, Z. Zhu, R. Owen, G. Jewell, and D. Howe, "Multiphase flux-switching permanent-magnet brushless machine for aerospace application," *Industry Applications, IEEE Transactions on*, vol. 45, no. 6, pp. 1971 –1981, 2009.

[77] D. Lin, P. Zhou, Z. Badics, W. Fu, Q. Chen, and Z. Cendes, "A new nonlinear anisotropic model for soft magnetic materials," *Magnetics, IEEE Transactions on*, vol. 42, no. 4, pp. 963 –966, 2006.

[78] W. T. Inc., "Magnetic properties of the neodymium sintered rare earth magnets," 2013. [Online]. Available: http://www.wenmag.com/materials/products/ pdf/neodymiuminfo.pdf

[79] D. Hanselman, *Brushless Permanent Magnet Motor Design*. Writers' Collective, 2003.

[80] F. Madauss, "Untersuchung der induktiven Kopplungen zwischen den Phasen einer permanentmagnetisch erregten polyphasigen elektrischen Maschine," Student Research Project, Brandenburg University of Technology, Cottbus, 2011.

[81] A. Taieb Brahimi, A. Foggia, and G. Meunier, "End winding reactance computation using a 3D finite element program," *Magnetics, IEEE Transactions on*, vol. 29, no. 2, pp. 1411 –1414, 1993.

[82] K.-C. Kim, D.-H. Koo, and J. Lee, "The study on the overhang coefficient for permanent magnet machine by experimental design method," *Magnetics, IEEE Transactions on*, vol. 43, no. 6, pp. 2483 –2485, 2007.

[83] P. Zhou, D. Lin, W. N. Fu, B. Ionescu, and Z. Cendes, "A general cosimulation approach for coupled field-circuit problems," *Magnetics, IEEE Transactions on*, vol. 42, no. 4, pp. 1051–1054, 2006.

[84] M. Jabbar, H. Phyu, Z. Liu, and C. Bi, "Modeling and numerical simulation of a brushless permanent-magnet DC motor in dynamic conditions by time-stepping technique," *Industry Applications, IEEE Transactions on*, vol. 40, no. 3, pp. 763–770, 2004.

[85] W. Hong, W. Lee, and B. Lee, "Dynamic simulation of brushless DC motor drives considering phase commutation for automotive applications," in *Electric Machines Drives Conference, 2007. IEMDC '07. IEEE International*, vol. 2, 2007, pp. 1377–1383.

[86] B. Kerdsup and N. Fuengwarodsakul, "Dynamic model of brushless DC drive using FE method based characteristics," in *Power Electronics and Drive Systems (PEDS), 2011 IEEE Ninth International Conference on*, 2011, pp. 66–71.

[87] O. Mohammed, S. Liu, and Z. Liu, "A phase variable model of brushless DC motors based on finite element analysis and its coupling with external circuits," *Magnetics, IEEE Transactions on*, vol. 41, no. 5, pp. 1576–1579, 2005.

[88] M. Fazil and K. Rajagopal, "Nonlinear dynamic modeling of a single-phase permanent-magnet brushless DC motor using 2-D static finite-element results," *Magnetics, IEEE Transactions on*, vol. 47, no. 4, pp. 781–786, 2011.

[89] Y. Lai, K. Lee, J. Tseng, Y. Chen, and T.-L. Hsiao, "Efficiency comparison of PWM-controlled and PAM-controlled sensorless BLDCM drives for refrigerator applications," in *Industry Applications Conference, 2007. 42nd IAS Annual Meeting. Conference Record of the 2007 IEEE*, 2007, pp. 268 –273.

[90] E. Bostanci, "Modeling and testing of a brushless DC machine drive for automobile propulsion applications," Master's thesis, RWTH Aachen University, 2010.

[91] E. Levi, "Multiphase electric machines for variable-speed applications," *Industrial Electronics, IEEE Transactions on*, vol. 55, no. 5, pp. 1893–1909, 2008.

[92] J. Ferreira, "Improved analytical modeling of conductive losses in magnetic components," *Power Electronics, IEEE Transactions on*, vol. 9, no. 1, pp. 127 –131, 1994.

[93] G. Bertotti, "General properties of power losses in soft ferromagnetic materials," *Magnetics, IEEE Transactions on*, vol. 24, no. 1, pp. 621 –630, 1988.

[94] D. Lin, P. Zhou, W. Fu, Z. Badics, and Z. Cendes, "A dynamic core loss model for soft ferromagnetic and power ferrite materials in transient finite element analysis," *Magnetics, IEEE Transactions on*, vol. 40, no. 2, pp. 1318 – 1321, 2004.

[95] F. Quattrone and R. Lorenz, "Dynamic modeling of losses in electrical machines for active loss control," in *Energy Conversion Congress and Exposition (ECCE), 2012 IEEE*, 2012, pp. 47 –52.

[96] J. Hendershot and T. Miller, *Design of Brushless Permanent-Magnet Machines*. Motor Design Books, LLC, 2010.

[97] S. Ruoho, E. Dlala, and A. Arkkio, "Comparison of demagnetization models for finite-element analysis of permanent-magnet synchronous machines," *Magnetics, IEEE Transactions on*, vol. 43, no. 11, pp. 3964 –3968, 2007.

[98] A. Fukuma, S. Kanazawa, D. Miyagi, and N. Takahashi, "Investigation of AC loss of permanent magnet of SPM motor considering hysteresis and eddy-current losses," *Magnetics, IEEE Transactions on*, vol. 41, no. 5, pp. 1964 – 1967, 2005.

[99] C. Zwyssig, M. Duerr, D. Hassler, and J. Kolar, "An ultra-high-speed, 500000 rpm, 1 kw electrical drive system," in *Power Conversion Conference - Nagoya, 2007. PCC '07*, 2007, pp. 1577–1583.

[100] L. Schwager, A. Tüysüz, C. Zwyssig, and J. Kolar, "Modeling and comparison of machine and converter losses for PWM and PAM in high-speed drives," in *Electrical Machines (ICEM), 2012 XXth International Conference on*, 2012, pp. 2441–2447.

[101] E. Spooner and B. Chalmers, "'TORUS': a slotless, toroidal-stator, permanent-magnet generator," *Electric Power Applications, IEE Proceedings B*, vol. 139, no. 6, pp. 497–506, 1992.

[102] Y. Li, J. Zhu, Q. Yang, Z. Lin, Y. Guo, and C. Zhang, "Study on rotational hysteresis and core loss under three-dimensional magnetization," *Magnetics, IEEE Transactions on*, vol. 47, no. 10, pp. 3520–3523, 2011.

[103] Y. Guo, J. Zhu, P. Watterson, and W. Wu, "Comparative study of 3-D flux electrical machines with soft magnetic composite cores," *Industry Applications, IEEE Transactions on*, vol. 39, no. 6, pp. 1696–1703, 2003.

[104] G. Klempner and I. Kerszenbaum, *Handbook of Large Turbo-Generator Operation and Maintenance*, ser. IEEE Press Series on Power Engineering. Wiley, 2009.

[105] L. Wang, F. Huo, W. Li, Y. Zhang, Q. Li, Y. Li, and C. Guan, "Influence of metal screen materials on 3-D electromagnetic field and eddy current loss in the end region of turbogenerator," *Magnetics, IEEE Transactions on*, vol. 49, no. 2, pp. 939–945, 2013.

[106] Infineon, *High Precision Hall Effect Switch TLE4946H*, Infineon Technologies Co., 2003. [Online]. Available: http://html.alldatasheet.com/html-pdf/80178/INFINEON/TLE4946H/2938/9/TLE4946H.html

[107] DANFYSIK, "ULTRASTAB 867-700PI precision current transducer," GMW Associates, 2007. [Online]. Available: http://www.gmw.com/electric_current/Danfysik/866_867/documents/867-700PI_Installation_Package.pdf

[108] Signaltec, *Installation Manual MCTS*, 2013. [Online]. Available: http://signaltec.de/shuntsnstuff/wp-content/uploads/2011/01/mcts-10-2010-en.pdf

[109] S. Messtechnik, "MR5-200C micro-ohmmeter," 2010. [Online]. Available: http://www.sanpaelektronik.com/teknik/MR%205-200C.pdf

[110] Hameg, *Programmable LCR Bridge HM8118 Manual*, HAMEG Instruments GmbH, 2008. [Online]. Available: http://www.datatec.de/shop/artikelpdf/hm8118_1_d.pdf

[111] F. Madauss, "Messverfahren zur Modellbildung einer elektrischen Maschine," Final Year Project, Brandenburg University of Technology, Cottbus, 2012.

[112] Phillips, *KTY83 series silicon temperature sensors*, Phillips Semiconductors, 2012. [Online]. Available: http://www.nxp.com/documents/data_sheet/KTY83_SER.pdf

[113] HBM, "Innovative digital torque transducer T12: Ultra high precision torque transducer for industrial use and test benches," 2013. [Online]. Available: http://www.hbm.com/en/menu/products/transducers-sensors/torque/t12

[114] Yokogawa, "WT3000 precision power analyser," Yokogawa Electric Corp., 2012. [Online]. Available: http://tmi.yokogawa.com/de